STATISTICAL MECHANICS

Statistical Mechanics
A survival guide

A. M. GLAZER
Department of Physics, University of Oxford

and

J. S. WARK
Department of Physics, University of Oxford

This book has been printed digitally and produced in a standard specification in order to ensure its continuing availability

OXFORD
UNIVERSITY PRESS

Great Clarendon Street, Oxford OX2 6DP
Oxford University Press is a department of the University of Oxford.
It furthers the University's objective of excellence in research, scholarship,
and education by publishing worldwide in

Oxford New York

Auckland   Cape Town   Dar es Salaam   Hong Kong   Karachi
Kuala Lumpur   Madrid   Melbourne   Mexico City   Nairobi
New Delhi   Shanghai   Taipei   Toronto
With offices in
Argentina   Austria   Brazil   Chile   Czech Republic   France   Greece
Guatemala   Hungary   Italy   Japan   South Korea   Poland   Portugal
Singapore   Switzerland   Thailand   Turkey   Ukraine   Vietnam

Oxford is a registered trade mark of Oxford University Press
in the UK and in certain other countries

Published in the United States
by Oxford University Press Inc., New York

© Oxford University Press, 2001

The moral rights of the author have been asserted

Database right Oxford University Press (maker)

Reprinted 2009

All rights reserved. No part of this publication may be reproduced,
stored in a retrieval system, or transmitted, in any form or by any means,
without the prior permission in writing of Oxford University Press,
or as expressly permitted by law, or under terms agreed with the appropriate
reprographics rights organization. Enquiries concerning reproduction
outside the scope of the above should be sent to the Rights Department,
Oxford University Press, at the address above

You must not circulate this book in any other binding or cover
And you must impose this same condition on any acquirer

ISBN 978-0-19-850816-8

To Pam and Tina

Preface

Statistical mechanics is the science of predicting the observable properties of a many-body system by studying the statistics of the behaviour of its individual constituents – be they atoms, molecules, photons etc. It provides the link between macroscopic and microscopic states. For example, it enables us to derive the pressure of an ideal gas (a macroscopic state) starting from the solutions to Schrödinger's equation for a particle in a box (the microscopic state). As such it has the potential to be one of the most satisfying parts of an undergraduate science course – linking in an elegant manner the quantum world with everyday observables of systems containing large numbers of particles.

However, in our experience, all too often students fail to appreciate the beauty of this bridge between worlds owing to the complexity of the mathematics with which they are presented. As with any subject, statistical mechanics contains its subtleties, nuances and insights that can only be fully understood after a thorough grounding in the underlying theory. However, the majority of the important, basic concepts are, in our opinion, readily accessible without resorting to such in-depth analysis. Therefore, right at the outset, it is important to state that the approach to the subject taken within this book is going to be slightly handwaving: we make no attempt to be sophisticated (just as well, some might say). We firmly believe that it is better to grasp the basic concepts, even if they are put forward in a non-rigorous way, than be presented with rigorous derivations but not understand them. For the brave-hearted, there are many more watertight approaches to the subject, to which we refer in the bibliography.

This book grew out of the experience of teaching statistical mechanics to physics undergraduates at the University of Oxford over many years. At present the Oxford undergraduate encounters this subject towards the end of the second year of a 3 – 4 year course, having previously received a course on classical equilibrium thermodynamics and an introduction to quantum mechanics. The material presented here covers the entire syllabus as it is taught at this level, and a more detailed approach (not included here) is then given for those working towards a more theoretical course. We supply a considerable number of questions, many of which have been traditionally asked at Oxford, together with worked solutions. The latter no doubt will not endear us to our colleagues, since they will now have to find new questions (without written answers) to set their students!

There are many approaches to studying this subject, and the most appropriate will clearly depend on your starting point. You will get the most out of this book if you have a basic knowledge of quantum mechanics (for example, the ability to solve Schrödinger's equation for simple one-dimensional potentials), though this is not essential. Furthermore, there is clearly a very strong connection between thermodynamics and statistical mechanics. We assume some prior study of thermodynamics, and thus at least a passing familiarity with the concept of entropy – though perhaps you may still find it somewhat confusing. It is a concept that often engenders a feeling of unease. If you have taken a thermodynamics course you will no doubt recall that there are elegant arguments proving that there is such a quantity called entropy and that it can be shown to be a function of state. But it would not be surprising if you remained somewhat confused as to its nature. The other aspect of thermodynamics that can be troubling, and which illustrates both its elegance and its apparent obscurity, is its complete generality. The derivation of all the TdS equations and suchlike is completely independent of the system in question – we can use them for ideal gases, magnets, and elastic bands – whatever we like. At no stage in a classical thermodynamic treatment do we worry about the detailed, nitty-gritty of the system of interest – its quantum states and the nature of the particles making up the system.

Now that's where statistical mechanics comes in. It will, we hope, give you a better feel for the nature of entropy, and provide an understanding of that important link between the microscopic and macroscopic world. By the end of this book you should be able to start from the microscopic world of Schrödinger's equation and derive the pressure (or entropy, enthalpy, etc.) of the gas in a steam engine, the magnetic susceptibility of a paramagnet, the heat capacity of a solid, and many, many more properties of a system of particles. It is worth realizing that many of the basic ideas in statistical mechanics were worked out in the 19th century by those giants of physics, Ludwig Boltzmann, Willard Gibbs, and James Clerk Maxwell. Boltzmann, in particular, found himself under serious attack by many of his colleagues for espousing the idea that the properties of gases and other systems could be treated in statistical terms. It is most interesting to read through those early papers and see the often tortuous and difficult arguments with which he had to wrestle, especially since today we think of many of the ideas in much simpler terms. In the event, Boltzmann was eventually shown to be correct. It was not until the 1920s that Heisenberg and others demonstrated the notion that the properties of quantum particles were indeed statistical in nature, rather than deterministic. Therefore, the work of these 19th century physicists laid the foundation for the modern science of quantum mechanics.

We are very grateful to Stephen Blundell, Dave Donaldson, and Anton Machacek, who took the time to read the draft manuscript of this book and made many helpful suggestions. Geoff Brooker made helpful comments on the original lecture notes upon which this book were based. Victor Zhitomirsky and Svetlana Zhitomirskaya produced the diagrams. Peter Sondhauss patiently answered our questions concerning LaTeX. We extend our thanks to them all.

Clarendon Laboratory, Oxford Mike Glazer
2001 Justin Wark

Contents

1	**Back to Basics**	1
1.1	The Role of Statistical Mechanics	1
1.2	Introduction to Coin Tossing	2
1.3	Heads I Win, Tails You Lose	2
1.4	Stirling's Theorem	5
1.5	More Flipping Coins	6
1.6	Distinguishable Particles	8
1.7	Summary	11
1.8	Problems	12
2	**The Statistics of Distinguishable Particles**	14
2.1	The Boltzmann Distribution for Distinguishable Particles	14
2.2	Lagrange's Method of Undetermined Multipliers	15
2.3	The Single-Particle Partition Function	21
2.4	Degeneracy	22
2.5	The Partition Function of a System	23
2.6	Summary	25
2.7	Problems	26
3	**Paramagnets and Oscillators**	29
3.1	A Spin-1/2 Paramagnet	29
3.2	Paramagnets with Angular Momentum **J**	34
3.3	The Simple Harmonic Oscillator	36
	3.3.1 An Array of 1-D Simple Harmonic Oscillators	36
	3.3.2 An Array of 3-D Simple Harmonic Oscillators	38
3.4	Summary	40
3.5	Problems	41
4	**Indistinguishable Particles and Monatomic Ideal Gases**	43
4.1	Distinguishable and Indistinguishable States	43
4.2	Identical Gas Particles – Counting the States	45
4.3	The Partition Function of a Monatomic Ideal Gas	51
4.4	Properties of the Monatomic Ideal Gas	52
4.5	More about Adiabatic Expansions	53

xii Contents

	4.6	Maxwell–Boltzmann Distribution of Speeds	56
	4.7	Gibbs Paradox	56
	4.8	Summary	59
	4.9	Problems	61
5	**Diatomic Ideal Gases**		**62**
	5.1	Other Degrees of Freedom	62
	5.2	Rotational Heat Capacities for Diatomic Gases	63
	5.3	The Vibrational Partition Function of a Diatomic Gas	65
	5.4	Putting it All Together for an Ideal Gas	66
	5.5	Summary	67
	5.6	Problems	68
6	**Quantum Statistics**		**69**
	6.1	Indistinguishable Particles and Quantum Statistics	69
	6.2	Bose–Einstein Statistics	71
	6.3	Fermi–Dirac Statistics	73
	6.4	More on the Quantum Distribution Functions	73
	6.5	Summary	76
	6.6	Problems	76
7	**Electrons in Metals**		**77**
	7.1	Fermi–Dirac Statistics: Electrons in Metals	77
	7.2	The Heat Capacity of a Fermi Gas	79
	7.3	The Quantum–Classical Transition	82
	7.4	Summary	85
	7.5	Problems	86
8	**Photons and Phonons**		**88**
	8.1	The Photon Gas	88
	8.2	Generalized Derivation of the Density of States	89
	8.3	Blackbody Radiation	90
	8.4	Phonons	92
	8.5	Summary	93
	8.6	Problems	94
9	**Bose–Einstein Condensation**		**97**
	9.1	Introduction	97
	9.2	The Phenomenon of Bose–Einstein Condensation	97
	9.3	The Quantum–Classical Transition Revisited	102
	9.4	Summary	102
	9.5	Problems	103
10	**Ensembles**		**104**
	10.1	Introduction	104
	10.2	The Chemical Potential	106

	10.3 Ensembles and Probabilities	107
	10.4 The Fermi–Dirac and Bose–Einstein Distributions Revisited	110
	10.5 Fermi–Dirac Distribution Revisited	110
	10.6 Bose–Einstein Distribution Revisited	111
	10.7 Summary	111
	10.8 Problems	112
11	**The End is in Sight**	**113**
	11.1 Phase Space	113
	11.2 Equipartition of Energy	115
	11.3 Route Map through Statistical Mechanics	116
	11.4 Summary	118
	11.5 Problems	118
A	**Worked Answers to Problems**	**119**
	A.1 Chapter 1	119
	A.2 Chapter 2	120
	A.3 Chapter 3	122
	A.4 Chapter 4	124
	A.5 Chapter 5	125
	A.6 Chapter 6	126
	A.7 Chapter 7	126
	A.8 Chapter 8	128
	A.9 Chapter 9	132
	A.10 Chapter 10	133
	A.11 Chapter 11	134
B	**Useful Integrals**	**137**
C	**Physical Constants**	**138**
D	**Bibliography**	**139**
Index		**140**

1
Back to Basics

Everything in the Universe is the fruit of chance and necessity.
Diogenes Laertius IX

1.1 The Role of Statistical Mechanics

In any undergraduate physics course, students will meet a certain number of 'core' subjects with which any self-respecting physicist must become familiar. Of these, this book relates to three core disciplines.

The first is what is termed *classical thermodynamics*. This is a subject dealing with the very large. It describes the world that we all see in our daily lives, knows nothing about atoms and molecules and other very small particles, but instead treats the universe as if it were made up of large-scale continua. It is therefore a science that considers everything in macroscopic terms. You should already be somewhat familiar with this subject by the time you read this book, and so you are already aware that thermodynamics describes such global quantities as heat, work, and the various functions of state, internal energy, entropy, enthalpy and free energies. Armed with these concepts, one is able to say a great deal about the behaviour of the world around us, whether it is simply a matter of heat engines and refrigerators, the properties of gases, or even fundamental conclusions about the universe. However, because it is a macroscopic subject, it lacks any perception of fine detail.

The second related subject is *quantum mechanics*. This is the other end of the spectrum from thermodynamics: it deals with the very small. It recognises that the universe is made up of particles: atoms, electrons, protons and so on. One of the key features of quantum mechanics, however, is that particle behaviour is not precisely determined (if it were, it would be possible to compute, at least in principle, all past and future behaviour of particles, such as might be expected in a classical view). Instead, the behaviour is described through the language of probabilities. This probabilistic approach is enshrined in the concept of the wave function, and armed with the famous Schrödinger equation a complete description of the probabilities should be possible. However, in practice, as everyone discovers, while an excellent solution to the hydrogen atom can be found (the observed energy of the ground state of -13.6 eV can be found easily with the appropriate wave function), not much else can without the help of

2 *Back to Basics*

enormous computation. When one brings into play more than one or two particles, the problem becomes rapidly insoluble, and so one has to resort to increasingly exotic approximations the more particles are brought into play. Imagine, therefore, trying to use quantum mechanics to describe the behaviour of 10^{30} atoms in a gas!

What we need therefore, if we are to reconcile the two opposite extremes of on the one hand a classical subject like thermodynamics and the non-classical quantum mechanics, is some means of bridging the two. This is where the third subject, *statistical mechanics*, comes to the rescue. Here, huge numbers of particles are considered within certain confined systems: no attempt is made at analysing the behaviour of individual particles, but instead, as in thermodynamics, one arrives at the overall macroscopic behaviour using probabilities. One therefore sees the natural link between the average behaviour of the thermodynamic world and the probabilistic world of individual particles. It was this linkage between the two that was the great triumph of the 19th century physicists Maxwell, Gibbs and, especially, Boltzmann. Well before quantum probabilistic ideas were applied to individual particles by Heisenberg in the 1920s, these gentlemen were already treating the world in probabilistic terms.

1.2 Introduction to Coin Tossing

As with any subject, a grasp of the basic concepts is absolutely essential and simple illustrations and toy problems can go a long way in aiding understanding. Therefore, rather than launch headlong into areas of real physical interest, let's start our study of the probabilistic nature of statistical mechanics by looking at a situation with which we are all familiar: flipping coins. This toy problem will illustrate the basic ideas, and introduce you in a non-threatening manner to some of the language of statistical mechanics. Of course, once we have dispensed with coins, we shall then expect to use some of the ideas generated and apply them to large (and we mean really massive) collections of very small particles like atoms and molecules. Although an atom is far from being equivalent to a coin (an atom doesn't have a head or a tail for example!) we shall see that nevertheless coins can teach us some useful ideas about statistics in general. And it is because in reality we want to deal with huge numbers of random particles that statistics plays such an important role.

1.3 Heads I Win, Tails You Lose

Let us assume that we are going to toss a coin 4 times. Let H denote a head, and T a tail. Here are a few possible outcomes for the series of 4 flips: $HHTH, THHT, THTH, TTTT$, etc., where $THHT$ means a tail on the first flip, a head on the second flip, and so on. There are obviously more possibilities than just those listed above. How many different ways in total are there of doing this? Well, we have 2 ways of tossing the coin on each of the 4 throws, so the total number of arrangements is $2^4 = 16$, and we have listed just 4 of them (chosen at random) above. Now, in the language of statistical mechanics, *each* of the specific arrangements ($HHTH, THHT, \ldots$) is called a *microstate*. We make a postulate:

- **Postulate 1: For a system in equilibrium all microstates are equally probable.**

What does this postulate mean for this coin problem? Well, it should be obvious that each time we flip the coin, we have a 50% chance of getting a head (or tail). So, we have just as much a chance of getting 4 heads in a row, $HHHH$, as getting the specific ordering $HTHT$. Each of these is a microstate, and each is as likely as any other. All the 16 possible microstates have the same likelihood of occurrence (obviously, each one has a 1 in 16 chance of occurring). It's a bit like a lottery. In the UK National Lottery you pick 6 different numbers from 1 to 49. People have various and curious ways of picking their numbers (the date of the cat's birthday, phase of the moon, number of their bank account, etc.) and hold the peculiar belief that some numbers are luckier than others. Interestingly, most people would not dare to pick the numbers, say, 1, 2, 3, 4, 5, and 6 as they have a 'gut' feeling that the lottery would never pick such a set of numbers. Yet this set is as likely as any other. This gives you some idea of how rare an individual microstate is! Each set of numbers is an equally-weighted 'microstate' (see Problem 1 in Section 1.8), and each is as likely to condemn you to 15 minutes of fame and jealous gloating by your former friends as any other selection. Needless to say, neither of us is in this position!

So we have learnt what a microstate is (at least for coins). What next? Well, we intuitively know something else about the outcome. We know that, whatever the *detailed* arrangement, when we flip the coins, the most likely thing to happen *overall* is that we will get 2 heads and 2 tails. The number of heads and tails, regardless of the order in which they come, would be called the *macrostate*. The macrostate is a macroscopic state of the system in which we are really interested. For coins, we are usually only interested in the total number of heads, and total number of tails, rather than the detail. That is to say, we are concerned whether we get 2 heads and 2 tails, rather than whether the specific order of the flips was $HHTT$ or, for example $HTHT$.

- **Postulate 2: The observed macrostate is the one with the most microstates.**

We said above that we instinctively know that the most likely outcome is to get 2 heads and 2 tails. From a statistical point of view, this is because it has more microstates than any other macrostate. To be specific, let us first look at the macrostate which gets 4 heads. There is only 1 microstate: $HHHH$. But the macrostate of 2 heads and 2 tails has 6 microstates: they are $HHTT, HTHT, HTTH, TTHH, THTH, THHT$. So we are 6 times as likely to get the macrostate of 2 heads and 2 tails than we are to get the

Table 1.1 The 5 macrostates for flipping a coin 4 times with the associated microstates

Macrostate	Microstates
4 Heads	HHHH
3 Heads and 1 Tail	HHHT
"	HHTH
"	HTHH
"	THHH
2 Heads and 2 Tails	HHTT
"	HTHT
"	THHT
"	HTTH
"	THTH
"	TTHH
1 Head and 3 Tails	TTTH
"	TTHT
"	THTT
"	HTTT
4 Tails	TTTT

macrostate of 4 heads. In total, there are 5 possible macrostates, and the microstates associated with each of these macrostates are shown in Table 1.1.

In this problem, we have the basic elements of statistical mechanics. We have learnt that a microstate is the detailed arrangement and each microstate is equally probable. However, the macrostate, which we will define to be the situation of macroscopic interest, can contain several microstates, and obviously the most probable macrostate is the one that contains the most microstates.

Now, in this simple example, it is clear that the most likely outcome is 2 heads and 2 tails – but not by much. We could have got 3 heads and 1 tail with a 4 in 16 chance, or even all heads (or tails) with a 1 in 16 chance. However, it is clear that when we come to deal with real physical systems, the numbers are going to get much larger than 4. We know that one gram mole of gas at standard temperature and pressure (STP) contains 6×10^{23} (Avogadro's number) molecules. When we come to deal with such systems we will find that the probability for the most likely macrostate is far more tightly 'peaked' than in the example involving 4 coins above (as an exercise, work out the numbers for tossing 6 coins). It is because the distribution becomes so tightly peaked that we make the second postulate above, that the macrostate observed *is* the one with the most microstates. If we had many, many coins, the probability becomes so narrowly peaked around half the number of heads and half the number of tails that to say this is the observed macrostate does not introduce a large error.

Also, it should be clear by now that within statistical mechanics we will often be interested in the factorial of a number. We can see this by thinking a little more generally about the coins. Let's flip a coin N times, and denote by the symbol Ω the number of ways of getting n heads (and therefore $(N-n)$ tails). We find

$$\Omega = \frac{N!}{n!(N-n)!} \quad . \tag{1.1}$$

It is important to understand this equation. If we have N different objects, we can arrange them in $N!$ different ways. However, when we flip a coin N times, and get n heads, then each result of heads looks the same: if you think of the result $HTHH$, it looks the same if we interchange any of the heads with any other head. Therefore, the number of ways of arranging the N flips is $N!$ divided by $n!$ (to take into account that all the heads look the same), and obviously divided by $(N-n)!$ to take into account that all the tails are the same also.

Notice how this fits in with flipping the coins 4 times. In this case, to get 2 heads (and therefore 2 tails) $\Omega = 4!/(2! \times 2!) = 6$ as expected. When we get to real physical problems we will thus be dealing with numbers like $10^{23}!$, and before proceeding further we are going to make a slight diversion to the topic of Stirling's theorem, which gives us a handy way to deal with $n!$ when n is extremely large. We will then return to the coins.

1.4 Stirling's Theorem

Stirling's theorem is an extremely useful formula for $n!$ that we will use again and again in our subsequent analysis of microstates. The simplest form of Stirling's theorem states that for large n

$$\ln n! \simeq n \ln n - n \quad .$$

You may wish to test the accuracy of this formula using a calculator (see Problem 5 in Section 1.8). This remarkable formula can be justified by a graphical method. Study Fig. 1.1. The dotted line is the function $y = \ln x$, where x is a continuous variable. The area under the curve from $x = 1$ to $x = n$ is the integral, I:

$$I = \int_1^n \ln x \, dx = \left[x \ln x - x \right]_1^n = n \ln n - n + 1 \quad .$$

6 Back to Basics

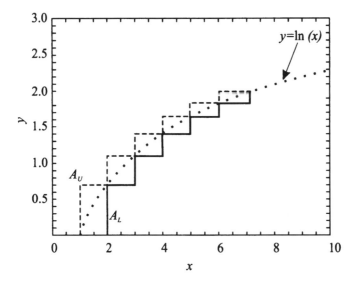

Fig. 1.1 Graphical proof of Stirling's theorem – see the text for an explanation.

We can obviously ignore the 1 for very large n. Also shown on the figure are two staircases, one above (the dashed line) and one below (the solid line) the smooth curve. The area under the upper one, A_U, between $x = 1$ and $x = n$ is the series:

$$A_U = \ln 2 + \ln 3 + \ln 4 + \ldots\ldots + \ln n = \ln n!$$

(as $\ln 1 = 0$ we don't need to include it). On the other hand between the same limits for the lower staircase:

$$A_L = \ln 2 + \ln 3 + \ln 4 + \ldots\ldots + \ln(n-1) = \ln(n-1)! = \ln n! - \ln n \quad .$$

So $A_L = \ln n! - \ln n$ and $A_U = \ln n!$. But we know the area under the line that passes right between them (the integral, $n \ln n - n$). From this we can deduce Stirling's theorem:

$$\ln n! = n \ln n - n \quad ,$$

with an error term of order $(\ln n)/2$.

The importance of this result is that we shall find later that the quantity $\ln \Omega$ plays a very important role in statistical mechanics. This is most fortunate because using logarithms changes huge numbers into relatively small, more tractable numbers. Thank goodness for logarithms!

1.5 More Flipping Coins

Armed with Stirling's theorem let us go back to flipping coins. We know intuitively, as we have said before, that the most probable result is evidently

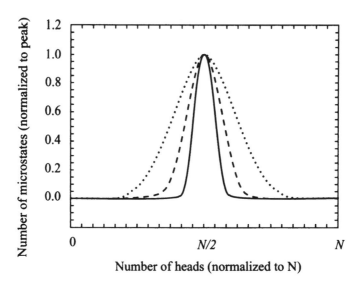

Fig. 1.2 The width of the distribution of the number of microstates per macrostate narrows as the number of coins increases: (i) dotted – about 20 coins, (ii) dashed – about 75 coins, (iii) solid – about 150 coins.

$n = N/2$ – that is to say, half heads and half tails. For the case of 4 coins we showed this explicitly by working out the number of ways it could occur, and we satisfied ourselves that this was the most likely outcome. But how do we prove this for an arbitrary number of coins, especially if the number of flips is very large? Well, if n is very large, we can treat it as a continuous, rather than discrete, variable, and we can work out where the maximum in Ω is by differentiating it with respect to n, and setting the derivative to zero. However, rather than differentiate Ω itself with respect to n, we will differentiate $\ln \Omega$ with respect to n, as this must also maximise at the same value of n, and it will turn out that this is far easier to deal with if we use good old Stirling's theorem. So, from eqn (1.1), we get

$$\ln \Omega = \ln N! - \ln n! - \ln(N - n)! \quad ,$$

which with the help of Stirling's theorem becomes

$$\ln \Omega = N \ln N - N - n \ln n + n - (N - n)\ln(N - n) + (N - n) \quad .$$

For a given total number of flips (i.e. keeping N constant) it is trivial to show that

$$\frac{d(\ln \Omega)}{dn} = -\ln n + \ln(N - n) = \ln(\frac{N - n}{n}) \quad ,$$

which is zero when $(N - n) = n$, and thus $n = N/2$ as expected.

We should note that although we have now proved that the most probable distribution is half heads and half tails, we have still not worked out

8 *Back to Basics*

how sharp the peak is. This point is dealt with in part in Problem 1 in Section 2.7. Here, we simply quote the result: for coins the width of the peak scales as $N^{1/2}$. But the total number of particles scales as N, so the *fractional* width scales as $N^{-1/2}$. This is best illustrated by Fig. 1.2, where the distribution (normalized to the peak) is shown for various numbers of coins. Similarly, we shall assume in the physics that follows that for systems containing large numbers of particles the fluctuations from the most probable result are limitingly small, and we are justified in taking the most probable macrostate (the one with the most microstates) as representing the so-called equilibrium state of the system.

1.6 Distinguishable Particles

Having learnt some of the basics by considering coins, we will now go on to look at situations of greater physical significance. We will start by looking at the statistics of distinguishable particles, i.e particles to which we can attach labels and identify them individually. For example, we can clearly tell the difference between atoms of different elements: if we have a helium atom and a neon atom in a box, and we walk away and return an hour later, we would still be able to tell them apart. We can also tell the difference between atoms of the same element, as long as they are localized in space. Consider Fig. 1.3. Here we have a set of atoms arranged to form a crystal. The atoms might be the same element, but each is still distinguishable from its neighbours because of its position. We might pick a particular atom – say the one marked by the arrow in the diagram and then walk away. When we return to look at the crystal, we will still be able to distinguish that particular atom from all the rest, because it will still be in the same place.

We will assume that each particle can occupy discrete energy levels (like quantum levels – so you can see one way in which quantum mechanics is going to enter straight away). Furthermore, to make things easier at this stage, we will assume that the levels are non-degenerate. That is to say, for the moment we consider that each energy level represents a different quantum state, and thus is statistically weighted by the same amount (we will deal with degeneracy in Section 2.4). So, for the crystal we treat each atom as though it were in its own little potential well which quantizes the energies. Note that at this stage we haven't said anything about the shape of the well, and therefore we don't know anything about the spectrum of the energy levels. All we know is that there is a spectrum of energy levels, and this spectrum is the same for each particle, because the shape of the potential well is the same for each of them. We say that the energy of the first level is ε_1, of the second level is ε_2, and so on.

Suppose there are a total of N particles in our system of particles, and we give a total energy U to the system. We will assume our system is isolated from the rest of the environment, so that no energy can escape. After giving this energy to the system, it will be distributed amongst the particles. Strictly speaking, for this to happen there must be some interaction between the particles so that energy can be transferred between them – we will assume that the interaction between the particles is just strong enough for this transfer to take place, but not sufficiently strong to prevent us from treating each particle independently. In order to accommodate this energy, some of the particles, let's say n_1 of them, will be in the first energy level of their particular potential wells, each with energy ε_1, some of them, n_2, will be in the second energy level ε_2, and so on. In general there will be n_i particles each with energy ε_i. Clearly, as we have a fixed number of particles:

$$\sum_i n_i = N$$

(we didn't write down such an equation for the coins, as it was clear that if we had n heads we must have $(N - n)$ tails). In addition, we have a further constraint that didn't bother us with the coins. As we have defined our system as being isolated, the total energy of the system, U, must be constant. That is to say

$$U = \sum_i n_i \varepsilon_i \quad.$$

So, we have $(n_1, n_2, n_3, ..., n_i, ...)$ particles in each of the energy levels $(\varepsilon_1, \varepsilon_2, \varepsilon_3,$
$..., \varepsilon_i, ...)$. This distribution (which tells us how many particles there are in each energy level) is now our macrostate. Knowing how many particles

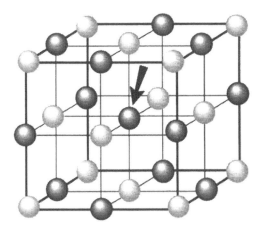

Fig. 1.3 Localized particles in a crystal are distinguishable – we can always identify the particle indicated by the arrow simply from its position.

there are with a particular energy is analogous to knowing how many heads and tails we are likely to get when we flip the coin a set number of times. With the coins we could only get heads or tails – now there are lots more possibilities – but the basic idea remains the same. For the particles, a particular microstate would entail knowing exactly which particular particles had which particular energies, rather than just knowing the overall numbers – just as for the coins a microstate entailed knowing exactly which flips gave the heads, and which ones the tails. For the coins we know that the macrostate of half heads and half tails is the most likely, as it contains the most microstates. In the same way, we need to find out the number of microstates in a macrostate for the particles. Once we have a formula telling us the number of microstates in a macrostate, we can find the one that has the maximum probability just as we did for the coins.

For the particles, the total number of microstates in a given macrostate, Ω, is given by

$$\Omega = \frac{N!}{n_1! n_2! n_3! \ldots n_i! \ldots} = \frac{N!}{\prod_i n_i!} \quad . \tag{1.2}$$

Notice how similar this is to the formula for the coins (eqn (1.1)). Instead of having just heads or tails (which for the coins meant that we divided the $N!$ by $n!$ and $(N-n)!$), we now have all the different possible energies $\varepsilon_1, \varepsilon_2, \varepsilon_3$ etc., containing n_1, n_2, n_3 particles, so we divide the $N!$ by $n_1!, n_2!, n_3!$ etc.

Before we wade further into the algebra, let us look at a simple example that sheds light on equation (1.2). Let us suppose that we have 7 localized particles, with energy levels that are equally spaced with energies $0, \varepsilon, 2\varepsilon, 3\varepsilon$ etc., such as occurs in simple harmonic oscillators, and we give our system of particles a total energy 4ε. Fig. 1.4 illustrates two possible microstates (there are many, many more in this system). For instance, (i) is a microstate where there are 6 particles in one level and 1 particular one in another. How many different ways in total can this be done? Answer = 7 (we could put any of the particles in the upper state). That is, there are 7 microstates in the corresponding macrostate. As for the microstate shown in (ii), this particular microstate is one of a total of 105 in the same macrostate! The macrostate corresponds to having 4 particles in the ground state, 2 in the level with energy ε, and 1 in the level with energy 2ε, and of course the microstate shown tells you exactly which of the seven particles has which energy. There are 105 microstates in this macrostate, as $\Omega = 7!/(4! \times 2! \times 1!) = 105$. One thing you might notice immediately is that the more probable distribution (the one with 105 microstates) looks as if it could be falling roughly exponentially with energy (4 particles in the ground state, 2 in the next, and 1 in the following state). This exponential fall-off – the so-called Boltzmann distribution law – is one of the main results of statistical mechanics. This law will be justified in Chapter 2. It so happens that this macrostate is the most probable one for this particular problem,

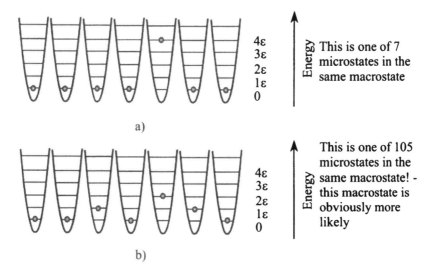

Fig. 1.4 Two particular arrangements (microstates) of 7 simple harmonic oscillators with a total energy of 4ε (we have neglected the zero point energy). In (a) all of the energy is taken up by one particular particle, whereas in (b) two particular particles each have energies of 1ε, while another particular particle has an energy of 2ε. The arrangement shown in (a) is one of 7 microstates in the same macrostate, whereas the arrangement shown in (b) is one of 105 microstates in the same macrostate.

just as getting half heads and half tails is the most probable macrostate for flipping coins. Obviously, just as with the coins, we would like to find a general way of finding the macrostate which corresponds to the maximum value Ω for the general case where N is very large. How this is done will be revealed in the next chapter.

1.7 Summary

To summarize the important points of this chapter:

- A microstate describes which particles are in which state. All allowed microstates are assumed equally probable.
- A macrostate describes how many particles are in each energy level. We assume that the observed macrostate is the one with the most microstates.
- Stirling's formula (in the form we shall use) is

$$\ln n! = n \ln n - n \quad .$$

- If we have N distinguishable non-degenerate particles, such that there are n_i particles in the energy level ε_i, the number of arrangements (i.e. the number of microstates per macrostate) is

$$\Omega = \frac{N!}{n_1! n_2! n_3! \ldots n_i! \ldots} = \frac{N!}{\prod_i n_i!} \ .$$

- For an isolated system of distinguishable particles, the number of particles and the energy are constant:

$$\sum_i n_i = N \ , \quad U = \sum_i n_i \varepsilon_i \ .$$

1.8 Problems

1. To win the UK National Lottery one must supposedly pick 6 different numbers from the 49 available. The order in which the numbers are chosen does not matter. If, being mean, we buy only one ticket, what are our chances of winning the jackpot?
2. (i) Mr. and Mrs. Fatchance have two children. If they tell you that at least one of them is a boy, but you do not know the gender of the other child, what is the probability that they have two boys?
(ii) If they had three children and told you again that at least one was a boy, what would the probability be that they were all boys?
(iii) Finally, suppose that they had never worked out what was causing all this procreation, and had N children, and told you again that at least one was a boy, what would the probability be that they were all boys?
3. In a particularly hideous game show, a contestant is asked to guess behind which of three closed doors a prize is hidden. He/she chooses one of the three doors, and tells the compère his/her choice. That door remains closed, but the compère (who knows where the prize actually lies) then opens one of the two remaining doors, showing it to be empty to the contestant. With the door of his/her original guess still closed, the contestant is then given the chance to guess again. Is there any advantage at this point to him/her in changing doors?
4. If we assume that the birthdays of the population are equally distributed throughout the year, how many people, taken at random, need to gather together in a room to give a greater than 50% chance of at least two of them sharing a common birthday?
5. Use your calculator to work out ln 10! Compare your answer with the simple version of Stirling's theorem $(N \ln N - N)$. How big must N be for the simple version of Stirling's theorem to be correct to within (i) 5% (ii) 1% ?
6. In Section 1.3 we worked out the number of ways we could get heads or tails when we flipped 4 coins (there were 6 ways of getting 2 heads and 2 tails, 4 ways of getting 1 head and 3 tails etc. etc.). Using eqn (1.1), work out the number of ways of getting 3 heads and 3 tails

when tossing 6 coins, and the number of ways of getting 2 heads and 4 tails, and 1 head and 5 tails, and so forth. Do the same for 10 coins. Plot a graph of your results, with the probability normalized to the peak of the distribution in each case, and with the number of particles normalized to the total number. Convince yourself that, in fractional terms, the distribution is getting slightly more sharply peaked as the number of coins increases.

2
The Statistics of Distinguishable Particles

He uses statistics as a drunken man uses lamp-posts – for support rather than illumination.
A. Lang

2.1 The Boltzmann Distribution for Distinguishable Particles

In Section 1.6 we showed that for a system of N particles the macrostate which has, in general, n_i particles in the non-degenerate energy level ε_i contains Ω microstates:

$$\Omega = \frac{N!}{\prod_i n_i!} \qquad (2.1)$$

But we also know that the total number of particles is constant, and, as we defined the system as being isolated from the environment, so too is the energy. That is to say

$$N = \sum_i n_i \quad \text{and} \quad U = \sum_i n_i \varepsilon_i \quad , \qquad (2.2)$$

where U is the total internal energy of the system. We now wish to maximise Ω to find out the most probable macrostate. Our approach is going to be very similar to that for the coins. First of all, we decide to maximise $\ln \Omega$ rather than Ω itself, for the simple reason that the mathematics is made a lot easier with the help of Stirling's theorem. So, taking logarithms of eqn (2.1):

$$\ln \Omega = \ln N! - \sum_i \ln n_i!$$

Then using Stirling's theorem we obtain

$$\ln \Omega = N \ln N - N - \sum_i n_i \ln n_i + \sum_i n_i = N \ln N - \sum_i n_i \ln n_i \quad . \quad (2.3)$$

In order to find the equilibrium state (the observed macrostate) we must maximize this quantity. For the coins maximizing Ω was fairly trivial – we just needed to differentiate with respect to the number of heads, n. Clearly, for the particles it is not quite as simple as this, as we can vary the number of particles in each of the energy levels, i.e. we can vary n_1, or n_2, or the number of particles in any level. Therefore, the maximum value of Ω will occur when

$$d(\ln \Omega) = \frac{\partial \ln \Omega}{\partial n_1} dn_1 + \frac{\partial \ln \Omega}{\partial n_2} dn_2 + \ldots \frac{\partial \ln \Omega}{\partial n_i} dn_i + \ldots = 0 \ .$$

But from eqn (2.3)

$$\frac{\partial \ln \Omega}{\partial n_i} = -(\ln n_i + 1)$$

and so

$$d(\ln \Omega) = - \sum_i (\ln n_i + 1) dn_i = 0 \ . \qquad (2.4)$$

But recall that the total number of particles, and the total internal energy are also constant, and so it follows that

$$\sum_i dn_i = 0 \quad \text{and} \quad \sum_i \varepsilon_i dn_i = 0 \ . \qquad (2.5)$$

Also, because of eqn (2.5), eqn (2.4) reduces to

$$d(\ln \Omega) = - \sum_i \ln n_i dn_i = 0 \ . \qquad (2.6)$$

2.2 Lagrange's Method of Undetermined Multipliers

It can be seen from the working above that in order to maximize $\ln \Omega$ (and hence Ω) we have three conditions to satisfy simultaneously:

$$\sum_i \ln n_i dn_i = 0, \quad \sum_i dn_i = 0, \quad \text{and} \quad \sum_i \varepsilon_i dn_i = 0 \ .$$

Physically, this corresponds to finding the macrostate with the maximum number of microstates subject to the constraints of a given number of particles and a given energy. As we have three quantities that must all be zero at the same time, we could certainly multiply two of them by arbitrary constants, say α and β:

$$\sum_i \ln n_i dn_i = 0, \quad \alpha \sum_i dn_i = 0, \quad \text{and} \quad \beta \sum_i \varepsilon_i dn_i = 0 \ .$$

The two constants, α and β, are known as Lagrange multipliers. We can now write

$$\sum_i dn_i (\ln n_i + \alpha + \beta \varepsilon_i) = 0 \ . \qquad (2.7)$$

How do we satisfy eqn (2.7)? As we have two unknowns, α and β, we might argue that we could satisfy two of the terms simultaneously, i.e. we could

16 The Statistics of Distinguishable Particles

eliminate the terms in dn_1 and dn_2 by satisfying the two simultaneous equations

$$\ln n_1 + \alpha + \beta\varepsilon_1 = 0 \quad \text{and} \quad \ln n_2 + \alpha + \beta\varepsilon_2 = 0 \ ,$$

but even if we did this it would still leave all the other terms in the series totally independent. Furthermore, there is no particular reason for picking the first and second terms in the series - we could pick any two we liked, and still leave all the rest independent. The result of this argument is that the only way to satisfy eqn (2.7) is to treat *all* of the terms as being mutually independent, and therefore each term in the series must be exactly zero, that is to say for all values of i

$$(\ln n_i + \alpha + \beta\varepsilon_i) = 0 \ ,$$

from which it follows that

$$n_i = \exp(-\alpha - \beta\varepsilon_i) = A\exp(-\beta\varepsilon_i) \ , \qquad (2.8)$$

where for convenience we have written $A = \exp(-\alpha)$.

This is starting to look a little like the expected Boltzmann distribution – the number of particles in a given energy level falls off exponentially with increasing energy. However, we are still left with the problem of determining the constants A and β. Note that if we knew what β was, then finding A would be trivial. A is just a constant of normalization, which can be seen by substituting eqn (2.8) into eqn (2.2):

$$N = A \sum_i \exp(-\beta\varepsilon_i) \ .$$

But how do we show what β is? Well, first, we will demonstrate that it must be some function of temperature.

Consider two systems separated by a thermally conducting wall as shown in Fig. 2.1. The number of particles in each of the two boxes is fixed as N_1 and N_2 respectively. The particles in the first box have a spectrum of available energy levels $\varepsilon_1, \varepsilon_2, \ldots \varepsilon_i \ldots$, with n_i particles in the energy level ε_i, whereas for the sake of generality we will allow the particles in the second box to have their own spectrum of available energy levels (not necessarily the same spectrum as for the first box), with m_i particles occupying the energy level ε'_i. As heat can flow from one box to the other, all we know about the energy is that the combined energy of the two systems, $(U_1 + U_2)$, is constant.

Now, as the total probability is the product of the partial probabilities, the number of microstates in a macrostate of the two systems together is clearly

$$\Omega = \frac{N_1!}{\prod_i n_i!} \times \frac{N_2!}{\prod_i m_i!} \ .$$

Lagrange's Method of Undetermined Multipliers 17

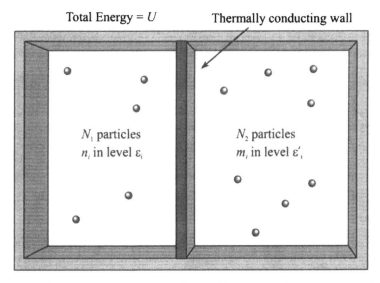

Fig. 2.1 Two separate systems of distinguishable non-degenerate particles are in thermal contact. Their combined total energy, $U = U_1 + U_2$, is constant.

We now crank the handle, and put this through the sausage machine once more (take logarithms, simplify using Stirling's theorem, and set the differential of $\ln \Omega$ equal to 0 to find the maximum):

$$\ln \Omega = N_1 \ln N_1 - \sum_i n_i \ln n_i + N_2 \ln N_2 - \sum_i m_i \ln m_i \quad,$$

$$d(\ln \Omega) = -d(\sum_i n_i \ln n_i) - d(\sum_i m_i \ln m_i) = 0 \quad,$$

$$\sum_i \ln n_i dn_i + \sum_i \ln m_i dm_i = 0 \quad, \tag{2.9}$$

where we have made use of the fact that the total number of particles in each box is fixed. This is now our condition for equilibrium (i.e. maximizing the total number of microstates in the macrostate). As the combined energy of the two systems is also constant, using Lagrange multipliers once again we conclude that eqn (2.9) must hold alongside the conditions:

$$\alpha_1 \sum_i dn_i = 0, \quad \alpha_2 \sum_i dm_i = 0, \quad \text{and} \quad \beta[\sum_i \varepsilon_i dn_i + \sum_i \varepsilon'_i dm_i] = 0 \quad. \tag{2.10}$$

Combining eqns (2.9) and (2.10) we find:

$$\sum_i dn_i(\ln n_i + \alpha_1 + \beta \varepsilon_i) + \sum_i dm_i(\ln m_i + \alpha_2 + \beta \varepsilon'_i) = 0 \quad.$$

18 The Statistics of Distinguishable Particles

Thus, as each of the sums must be zero (there is no connection between the number of particles in each of the boxes: as particles are not allowed to go from one box to the other, the sums involving dm_i and dn_i must be independent), we obtain

$$n_i = A_1 \exp(-\beta \varepsilon_i), \qquad m_i = A_2 \exp(-\beta \varepsilon_i') \ .$$

The important point is that β is the same for both of the systems (this was always going to be the case from eqn (2.10) onwards) and it is independent of $N_1, N_2, \varepsilon_i, \varepsilon_i'$, etc. This implies, as we suspected, that it is a function of temperature, $\beta = \beta(T)$, as we know from thermodynamics that this is the parameter that is the same for two systems in thermal contact and in equilibrium. The zeroth law of thermodynamics states that for two systems in thermal contact, energy will flow from one to the other until they reach equilibrium, at which point they have the same temperature. However, so far we have only shown that β is *some* function of temperature – the exact form of the function still remains unknown.

In order to demonstrate what that function actually is, we are going to have to make a postulate, as Boltzmann did, that the entropy, S, of the system is related to the number of microstates by the formula:

$$S = k_B \ln \Omega \ , \qquad (2.11)$$

where k_B is Boltzmann's constant ($k_B = 1.3807 \times 10^{-23}\,\text{J}\,\text{K}^{-1}$). Before we work out how this helps us to derive $\beta(T)$, a word about the above formula. It must be stressed that it is a postulate, indeed the most important postulate of statistical mechanics (it is even engraved on Boltzmann's tombstone); nevertheless, it remains a postulate. As such, it is no more provable than the laws of thermodynamics, or the assumption that the wave equation always correctly predicts the behaviour of particles. It is a guess – a very intelligent guess – which can be justified ultimately because it works. That said, we should note that there are some good reasons for believing that the entropy of a system should be of this form. For instance, (i) we know from thermodynamics that as systems go towards equilibrium, the entropy is maximized, and we have already stated that the equilibrium macrostate is the one with the most microstates, and (ii) we also know from thermodynamics that S is additive, but the laws of probability say that Ω must be multiplicative – the above equation meets this criterion as well because of the logarithm. You can now appreciate why we maximised $\ln \Omega$ in Section 2.1. Not only does it make the mathematics easier, it also corresponds to maximizing the entropy.

So, having made this giant leap of faith, let us return to the matter of how it helps us find out about the form of $\beta(T)$. Well, from the first law of thermodynamics we know that the increase in internal energy of a system, dU is made up of the heat input dQ and the work done on the system, dW:

$$dU = đQ + đW . \tag{2.12}$$

For a reversible change,

$$đQ = TdS . \tag{2.13}$$

Armed with these two equations, we can make the link between the microscopic world of statistical mechanics and the macroscopic world of pressures, temperatures, enthalpies, etc. that we find in thermodynamics. To do this, we note that the statistical viewpoint tells us that a small change in energy is given by

$$dU = d\left(\sum_i \varepsilon_i n_i\right) = \sum_i \varepsilon_i dn_i + \sum_i n_i d\varepsilon_i . \tag{2.14}$$

Compare the thermodynamic definition of dU with the statistical one. A moment's thought should convince you that the two terms on the right-hand sides of eqns (2.12) and (2.14) match up. The first term in eqn (2.14) is the increase in energy owing to moving the particles amongst the unchanged energy levels – that is to say, just changing the arrangements of the particles: clearly this changes the number of microstates and hence the entropy. It is therefore related to dS, and thus the term for heat input in eqn (2.12). On the other hand, the second term on the right-hand side of eqn (2.14) keeps the number of particles, n_i, in a given energy level ε_i constant – but we change the energy of that level. However, since the populations of the levels remain the same, we can't have altered the number of ways we can arrange the particles, i.e. the number of microstates – so this term has nothing to do with an entropy change: it must correspond to the work term of eqn (2.12). Again, this makes sense intuitively. If we imagine doing, for example, PdV type work on a system, say compressing a crystal, then when we pushed the atoms closer together we would alter the shape of the potential wells in which they were sitting and therefore change the spectrum of the energy levels. So the heat term corresponds to a disordering of the system, while the work term is an ordered change.

Now consider a small energy change that involves no work, and equate the first terms on the right-hand sides of eqns (2.12) and (2.14):

$$TdS = \sum_i \varepsilon_i dn_i . \tag{2.15}$$

Substituting eqn (2.11) into eqn (2.15) we find

$$Td(k_B \ln \Omega) = \sum_i \varepsilon_i dn_i ,$$

20 The Statistics of Distinguishable Particles

which, because of eqn (2.6), is equivalent to

$$-k_B T \sum_i \ln n_i \, dn_i = \sum_i \varepsilon_i \, dn_i \quad . \tag{2.16}$$

Substituting eqn (2.8) into eqn (2.16) yields

$$-k_B T \sum_i (-\alpha - \beta \varepsilon_i) dn_i = \sum_i \varepsilon_i \, dn_i \quad .$$

The term in α clearly sums to zero (because of the constant number of particles), and we finally deduce the result we were after:

$$\beta = \frac{1}{k_B T} \quad . \tag{2.17}$$

We find that β is indeed solely a function of temperature - it is inversely proportional to it. Note that by assuming $S = k_B \ln \Omega$ we found that $\beta = 1/k_B T$. Alternatively we could assume the second result to find the first. So our analysis depends on at least one unprovable assumption.

Having made this link between statistical mechanics and thermodynamics, we are now in a position to derive a complete list of equations relating thermodynamic functions to microscopic energies! Using eqns (2.2) and (2.8) these are:

$$n_i = A \exp(-\varepsilon_i/k_B T) \quad ,$$

$$N = A \sum_i \exp(-\varepsilon_i/k_B T) \quad , \tag{2.18}$$

$$U = A \sum_i \varepsilon_i \exp(-\varepsilon_i/k_B T) \quad . \tag{2.19}$$

We can also find an expression for the entropy:

$$S = k_B \ln \Omega = k_B (N \ln N - \sum_i n_i \ln n_i) \quad .$$

Now, using the fact that $n_i = A \exp(-\varepsilon_i/k_B T)$, we have

$$\begin{aligned} S &= k_B (N \ln N - \sum_i n_i [\ln A - \varepsilon_i/k_B T]) \\ &= k_B (N \ln N - N \ln A + U/k_B T) \quad . \end{aligned} \tag{2.20}$$

But

$$N = \sum_i n_i = A \sum_i \exp(-\varepsilon_i/k_B T) \quad .$$

Therefore,
$$\ln N = \ln A + \ln \sum_i \exp(-\varepsilon_i/k_B T) \quad , \tag{2.21}$$

and combining eqns (2.20) and (2.21) we obtain

$$S = Nk_B \ln \sum_i \exp(-\varepsilon_i/k_B T) + \frac{U}{T} \quad . \tag{2.22}$$

We will also add one more to this list, for reasons that will become apparent later. From thermodynamics we know that the Helmholtz free energy, F, is given by $F = U - TS$, and so

$$F = -Nk_B T \ln \sum_i \exp(-\varepsilon_i/k_B T) \quad . \tag{2.23}$$

2.3 The Single-Particle Partition Function

If we look at the equations given above, we notice that the particular term $\sum_i \exp(-\varepsilon_i/k_B T)$ keeps turning up. We give it a special name – the Partition Function, denoted by the letter Z, which stands for the German 'Zustandsumme', meaning 'sum over states'. We define

$$Z = \sum_i \exp(-\varepsilon_i/k_B T) \quad .$$

Strictly speaking, the partition function we have written down is called the single-particle partition function, Z_{sp}, as it is independent of the total number of particles; more of this later. For now, let us note an extremely important point: if we know the partition function for a particular system, we then know *all* of the thermodynamic functions. It is difficult to overstress the importance of this. If you know the formula for the partition function you know everything about the thermodynamics of the system. Consider, to take an example, an ideal gas. If we knew the partition function of the gas, we could immediately deduce the equation of state and the heat capacity (we will actually do the mathematics of this in Chapter 4; just take our word for it now). Now, that may not seem such a big deal to you, but think about it for a second. Let's say we told you that we had an ideal gas, with equation of state $PV = RT$. Given that information, what is its heat capacity? You might say $3R/2$ (if you already knew the heat capacity of a monatomic gas). But hold on, we didn't tell you it was monatomic, only that it was ideal – so the heat capacity could be higher if it was diatomic or greater. That's the point – on its own the equation of state of a system tells you nothing about the heat capacity (and vice versa); on the other hand, as we shall show as we proceed, armed with the partition function of a system, we know everything about it: the equation of state, heat capacity, formula

22 The Statistics of Distinguishable Particles

for adiabatic expansion, etc, etc, etc. The partition function contains all of the thermodynamics. Whilst this is truly remarkable, there is a sense in which we should not be too surprised. In order to construct the partition function we need to know the quantum levels of the system in question – and clearly that piece of information must contain all of the pertinent physics. To put it another way, what else is there to know, apart from the spectrum of the energy levels?

Let's now write down our thermodynamic functions in terms of the single-particle partition function – these are extremely important, as they tell us how to get any thermodynamic potential from $Z_{\rm sp}$. We leave it as an exercise in algebra for you to demonstrate that eqns (2.18), (2.19), (2.22) and (2.23) can be written:

$$N = A Z_{\rm sp} \; , \tag{2.24}$$

$$U = N k_{\rm B} T^2 \frac{\partial \ln Z_{\rm sp}}{\partial T} \; , \tag{2.25}$$

$$S = N k_{\rm B} \ln Z_{\rm sp} + N k_{\rm B} T \frac{\partial \ln Z_{\rm sp}}{\partial T} \; , \tag{2.26}$$

$$F = -N k_{\rm B} T \ln Z_{\rm sp} \; . \tag{2.27}$$

Note the comparatively simple form of F in terms of $Z_{\rm sp}$. This is one of the beauties of F, the other main one being that it naturally leads us to the equation of state of the system, as we shall see shortly.

2.4 Degeneracy

Let us make our definition of the single-particle partition function a little bit more general. You will recall that when we started out, we made the assumption that each energy level was associated with a different quantum state, and on that basis was equally weighted. However, we know from our knowledge of quantum mechanics that situations can arise when different quantum states both have the same energy: this is known as degeneracy. How do we deal with this? Consider Fig. 2.2, where we show an arbitrary system which has three different energy levels. In Fig. 2.2(a) we assume that each energy level has one state associated with it – i.e. is non-degenerate. The partition function is clearly

$$Z_{\rm sp} = \exp(-\varepsilon_1/k_{\rm B}T) + \exp(-\varepsilon_2/k_{\rm B}T) + \exp(-\varepsilon_3/k_{\rm B}T) \; . \tag{2.28}$$

If we let the second and third energy levels get closer and closer together as shown in Fig. 2.2(b), they would eventually have the same energy, and it would look as though we only had two energy *levels*, but there would still be three quantum *states*. The upper level would need to be counted twice, as it has two states associated with it (it is now degenerate). This is the important point – it is the quantum states that are the entities that are given equal statistical weight, not the levels. So the partition function would, in this case where the upper state is now doubly degenerate and $\varepsilon_2 = \varepsilon_3$, be

$$Z_{sp} = \exp(-\varepsilon_1/k_B T) + 2\exp(-\varepsilon_2/k_B T) \quad . \tag{2.29}$$

In general, if the energy level ε_i has g_i quantum states associated with it (that is to say, it is 'g_i degenerate'), then the single-particle partition function is given by

$$Z_{sp} = \sum_i g_i \exp(-\varepsilon_i/k_B T) \quad , \tag{2.30}$$

where the sum is over the *levels*.

It is this form of Z_{sp} that should be used in eqns (2.24) through (2.27). It should also be obvious that, in the presence of degeneracy, the number of particles n_i with energy ε_i is now

$$n_i = A g_i \exp(-\varepsilon_i/k_B T) \quad .$$

2.5 The Partition Function of a System

Thus far we have considered the so-called single-particle partition function, defined as $Z_{sp} = \sum_i g_i \exp(-\varepsilon_i/k_B T)$, and then worked out thermodynamic potentials such as $F = -Nk_B T \ln Z_{sp}$. However, it is also possible to define a partition function for the whole system, which we denote by Z_N. We assume that the energy level occupied by each of the N particles in the

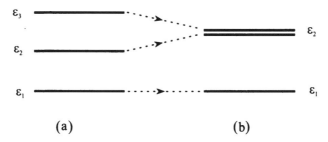

Fig. 2.2 (a) Three non-degenerate energy levels, giving rise to the partition function in eqn (2.28), (b) two of the quantum states now have the same energy, and thus if we sum over energy levels the partition function is now given by eqn (2.29).

system is independent of all the other particles, so that one particle is in, say, energy level ε_i (which is g_i degenerate), the next particle is in energy level ε_j (which is g_j degenerate), and so on. We can then think of the whole system of N particles being in some energy state, E_m, where

$$E_m = \varepsilon_i + \varepsilon_j + \varepsilon_k + \ldots ,$$

with a degeneracy g_m

$$g_m = g_i \times g_j \times g_k \times \ldots .$$

The partition function of the system is therefore

$$Z_N = \sum_m g_m \exp(-E_m/k_B T) ,$$

$$Z_N = \sum_i g_i \exp(-\varepsilon_i/k_B T) \times \sum_j g_j \exp(-\varepsilon_j/k_B T) \times \ldots = (Z_{\rm sp})^N ,$$

as there is a total of N terms.

We now can write $F = -k_B T \ln Z_N$. Similarly U and S are the same as before, apart from the factor of N. Thus, for distinguishable particles

$$Z_N = (Z_{\rm sp})^N , \tag{2.31}$$

$$U = k_B T^2 \frac{\partial \ln Z_N}{\partial T} , \tag{2.32}$$

$$S = k_B \ln Z_N + k_B T \frac{\partial \ln Z_N}{\partial T} , \tag{2.33}$$

$$F = -k_B T \ln Z_N . \tag{2.34}$$

You are probably asking yourself why we bothered to do this. What was wrong with the previous formulae: eqns (2.24) to (2.27)? Well, nothing really: they are totally equivalent to those given above. Whether we use the equations for the single-particle partition function or those for the partition function of the whole system, we will still obtain the same results. However, in many ways it is more useful to remember the equations for the partition function for a system. When we go on to deal with indistinguishable (rather than distinguishable) particles that obey Boltzmann statistics we will find that they also obey eqns (2.32) to (2.34), and thus we only need to remember one set of equations. However, for indistinguishable particles there is a different relationship between the single-particle partition function and that for the whole system – i.e. eqn (2.31) does not hold for the indistinguishable particles. This tasty morsel will be held over until Chapter 4.

Finally, even though we have assumed that each particle can occupy different quantized energy levels, this type of statistics is usually referred to as classical statistics. This admittedly somewhat confusing convention is used because even though the particles occupy quantized levels, the way in which we do the counting of the numbers of arrangements is purely classical, because we can tell the particles apart just as we can tell classical billiard balls apart.

2.6 Summary

- By maximizing the number of microstates, subject to the constraints of a given number of particles and a given total energy, we derived the Boltzmann distribution, which tells us how many particles are in a given energy level. Taking into account degeneracy:

$$n_i = Ag_i \exp(-\beta \varepsilon_i) \quad .$$

- Entropy is related to the number of microstates, Ω, by

$$S = k_B \ln \Omega \quad . \tag{2.35}$$

- With this definition of entropy and the thermodynamic definition of temperature we showed that $\beta = 1/k_B T$ and

$$n_i = Ag_i \exp(-\varepsilon_i/k_B T) \quad .$$

- We define a new quantity, called the single-particle partition function, $Z_{\rm sp}$, which is *very important*, as all the thermodynamic potentials can be derived from it. Again, taking into account degeneracy:

$$Z_{\rm sp} = \sum_i g_i \exp(-\varepsilon_i/k_B T) \quad .$$

- In terms of the single-particle partition function we have the following relations:

$$N = AZ_{\rm sp}$$

$$U = Nk_{\rm B}T^2 \frac{\partial \ln Z_{\rm sp}}{\partial T}$$

$$S = Nk_{\rm B} \ln Z_{\rm sp} + Nk_{\rm B}T \frac{\partial \ln Z_{\rm sp}}{\partial T}$$

$$F = -Nk_{\rm B}T \ln Z_{\rm sp}$$

- Alternatively, we can work in terms of the partition function of the *system*. For distinguishable particles the partition function of the system is related to the single-particle partition function by

$$Z_N = (Z_{\rm sp})^N \ .$$

Note that the above relation does *not* hold for indistinguishable particles. In terms of the partition function of the system we have the following relations:

$$U = k_{\rm B}T^2 \frac{\partial \ln Z_N}{\partial T}$$

$$S = k_{\rm B} \ln Z_N + k_{\rm B}T \frac{\partial \ln Z_N}{\partial T}$$

$$F = -k_{\rm B}T \ln Z_N \ .$$

Importantly, these three relations *do* still hold for indistinguishable particles obeying Boltzmann statistics, but for them we shall find a different relationship between the single-particle partition function and the partition function of the system.

2.7 Problems

1. If a coin is flipped N times, demonstrate that the number of ways, Ω, that one can get exactly half of them heads, and half of them tails is:

$$\frac{N!}{(\frac{N}{2}!)^2} \ .$$

In problem 6 in Chapter 1 you should have convinced yourself, just by working out the numbers, that the distribution was becoming slightly more and more peaked as the numbers of coins increased. Let us now prove this for a large number of flips by looking at the probability of getting nearly, but not quite, $N/2$ heads. Write down the number of ways in which one can get $(\frac{N}{2} - m)$ heads, and $(\frac{N}{2} + m)$ tails. Show that the number of ways of doing this tends to half of the value of getting exactly half heads and half tails when

$$\left(\frac{N}{2}-m\right)!\left(\frac{N}{2}+m\right)! = 2\left(\frac{N}{2}!\right)^2$$

and hence, assuming $N \gg m \gg 1$, show that

$$\left(\frac{N}{2}\right)^m + [1+2+3\ldots+m]\left(\frac{N}{2}\right)^{m-1}$$
$$\simeq 2\left[\left(\frac{N}{2}\right)^m - [1+2+3\ldots+m]\left(\frac{N}{2}\right)^{m-1}\right] \quad .$$

Noting that $[1+2+\ldots m \sim m^2/2]$ show that this leads to

$$m \sim \sqrt{N}$$

and that the fractional width therefore goes as $N^{-\frac{1}{2}}$. Comment on this result for $N \sim 10^{23}$.

2. In Section 1.6 we considered 7 localized particles, with equally spaced energy levels $0, \varepsilon, 2\varepsilon, 3\varepsilon$, etc., and we gave our system of particles a total energy 4ε. We looked at two possible macrostates. Now find all the possible macrostates for this system, along with their associated number of microstates. Furthermore, calculate the average population in each of the levels, assuming all microstates are equally probable. Compare this average with both the Boltzmann distribution and the distribution for the most probable macrostate. Comment on the difference.

3. Given that when we take into account degeneracy the total energy of the sytem is

$$U = A\sum_i g_i \varepsilon_i \exp(-\varepsilon_i/k_B T) \quad ,$$

where A is defined by normalization from

$$N = A\sum_i g_i \exp(-\varepsilon_i/k_B T) \quad ,$$

demonstrate that

$$U = Nk_B T^2 \left(\frac{\partial \ln Z_{\text{sp}}}{\partial T}\right) \quad ,$$

where

$$Z_{\text{sp}} = \sum_i g_i \exp(-\varepsilon_i/k_B T) \quad .$$

4. The separation of energy levels in atoms is typically of the order of a few electron volts. For an idealized two-level atom, with non-degenerate levels separated by 1 eV, calculate the ratio of the excited state to ground state populations at room temperature.

3
Paramagnets and Oscillators

I was at a loss as to what to do... I took stock of my qualifications. A not-very-good degree, redeemed somewhat by my achievements at the Admiralty. A knowledge of certain restricted parts of magnetism and hydrodynamics, neither of them subjects for which I felt the least bit of enthusiasm. No published papers at all... Only gradually did I realize that this lack of qualification could be an advantage.
F. Crick

3.1 A Spin-1/2 Paramagnet

Up until now everything has been quite general, and no doubt you are starting to question the relevance to the price of sugar of all these pages of formulae. So, before you get lost in the abstraction of it all, let us start to apply them to some relatively simple physical systems. We have picked two: a paramagnetic solid and an array of simple harmonic oscillators. We choose these examples as not only are they perhaps the simplest to deal with from a mathematical point of view, but they are also of considerable physical interest.

A crystal consists of a set of atoms, each in a particular position. The atoms are thus localized, and the statistics of distinguishable particles applies. In a paramagnetic crystal each atom has total angular momentum J. You may recall from your study of quantum mechanics that this means that the z component of the angular momentum can take on all the values from $-m_J$ to $+m_J$, and in the absence of a magnetic field the state is $(2J+1)$ degenerate. If the atom is in the particular quantum state such that it has a z component of angular momentum, m_J, then it will have an associated magnetic moment that is proportional to m_J (classically we think in terms of the electron with angular momentum 'orbiting' the nucleus, and thus looking like a miniature current loop that creates a magnetic moment). In the absence of a magnetic field, all of the states with different m_J have the same energy and thus are equally likely to be populated, and the net magnetic moment is zero. However, in the presence of a magnetic field, the states are shifted as described in more detail below, and the sample exhibits a magnetic moment along the direction of the field – this is known as paramagnetism.

30 Paramagnets and Oscillators

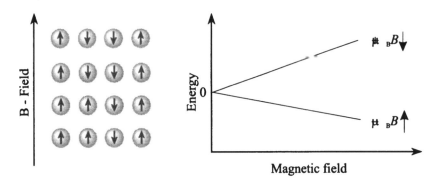

Fig. 3.1 The paramagnet is an array of particles with net angular momentum. We treat the simple case $J = S = 1/2$. In the presence of a magnetic field the spins align parallel or antiparallel with the field, with energies $\pm\mu_B B$.

For simplicity, we will start off by assuming $J = S = 1/2$, and so the state is only doubly degenerate. When we apply a magnetic field B, the degeneracy is lifted, and the two states shift in energy by $\pm g_J \mu_B m_J B$, where μ_B is the Bohr magneton, and g_J is the so-called Landé splitting factor, which is a number which relates the magnetic moment of an atom to its angular momentum. Be careful not to confuse it with the degeneracy g. For the simple spin-1/2 case, $m_J = \pm 1/2$ and $g_J = 2$, and so the energy levels split by $\pm\mu_B B$. The energy levels are shown in Fig. 3.1.[1]

Now, what's the single-particle partition function? Well, in this particular case it's easy, as there are only two levels (this is a bit like our first example of tossing coins, but now one of the states – the head or the tail, whichever we choose, has more energy than the other). Thus there are only two terms in the partition function:

$$Z_{sp} = \exp(-\mu_B B/k_B T) + \exp(+\mu_B B/k_B T)$$
$$= 2\cosh\left(\frac{\mu_B B}{k_B T}\right) \quad .$$

We have claimed that we can get all the thermodynamic functions from this partition function. As examples, we will derive the internal energy, the heat capacity, the entropy and the magnetization.

First the internal energy. Using eqns (2.32) and (2.31), we have

---

[1] Note that the magnetic moment due to the spin is antiparallel to the spin itself (i.e. the angular momentum) – the arrows in Fig. 3.1 correspond to the direction of the magnetic moment.

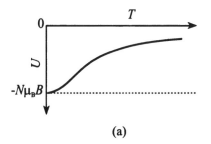 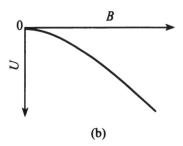

Fig. 3.2 (a) A schematic diagram of the internal energy as a function of temperature for a paramagnet in a fixed magnetic field, where zero energy is taken to be the energy of the system in the absence of the magnetic field. (b) Internal energy as a function of magnetic field at fixed temperature.

$$U = Nk_BT^2 \frac{\partial \ln Z_{sp}}{\partial T}$$
$$= Nk_BT^2 \frac{\partial}{\partial T}\left(\ln 2 + \ln \cosh \frac{\mu_B B}{k_B T}\right)$$
$$= -N\mu_B B \tanh\left(\frac{\mu_B B}{k_B T}\right) \quad . \tag{3.1}$$

It is important to note that this is the internal energy increase due to the paramagnet being placed in the magnetic field; a particular paramagnetic crystal will also have some internal energy due to the vibrations of its atoms, or other forms of internal energy – what we have calculated is only the extra internal enery due to the magnetic properties.

The internal energy as a function of T and B is shown in Fig. 3.2. Let us study this figure to see what we can learn. First of all, notice from Fig. 3.2(a) that at very low temperatures in the presence of a magnetic field, the energy of the paramagnet is $-N\mu_B B$. This is exactly what we would expect: the field splits the levels into two as shown in Fig. 3.1, but at very low temperatures all of the atoms must be in the ground state, each with energy $-\mu_B B$, and thus the total N atoms have an energy $-N\mu_B B$. As the system is heated up, some of the electrons are excited into the upper level, thus increasing the overall energy. At very high temperatures, such that the energy splitting between the levels is small compared with $k_B T$, the two levels become very nearly equally populated, and the original energy is obtained.

The internal energy as a function of B for a fixed temperature is shown in Fig. 3.2(b). Notice that at large magnetic fields the energy is proportional to $-B$. Again we find this unsurprising: at large magnetic fields the splitting between the two levels will be so large that there are very few atoms in the upper state even though the system is at finite temperature, and the energy will be $-N\mu_B B$ once more. Mathematically, this is because $\tanh x \to 1$ as $x \to \infty$, and we note that the argument of the tanh function in eqn (3.1) becomes very large at low temperatures. For weaker magnetic fields, the situation is slightly more complicated, for now the splitting between the levels is less, and some of the atoms are going to be thermally excited into the higher energy state. Indeed, for very weak fields the energy is proportional to $-B^2$, because for small x, $\tanh x \to x$.

The heat capacity can be found by differentiating the internal energy with respect to temperature:

$$C = \frac{dU}{dT} = -N\mu_B B \left(\frac{-\mu_B B}{k_B T^2}\right)\left[\cosh\left(\frac{\mu_B B}{k_B T}\right)\right]^{-2}.$$

Therefore

$$C = Nk_B \left(\frac{\theta}{2T}\right)^2 \frac{4}{(\exp(\theta/2T) + \exp(-\theta/2T))^2},$$

where

$$\theta = \frac{2\mu_B B}{k_B},$$

leading to

$$C = Nk_B \left(\frac{\theta}{T}\right)^2 \frac{\exp(\theta/T)}{(\exp(\theta/T) + 1)^2}.$$

The limiting forms of the heat capacity at high and low temperatures are

$$C = \frac{Nk_B}{4}\left(\frac{\theta}{T}\right)^2 \quad \text{as} \quad T \to \infty,$$

and

$$C = Nk_B \left(\frac{\theta}{T}\right)^2 \exp(-\theta/T) \quad \text{as} \quad T \to 0.$$

The heat capacity as a function of temperature is plotted in Fig. 3.3. Note that we predict a peak in the heat capacity due to the magnetic spins which depends on the applied field, and this peak occurs roughly when the temperature is of the order of $\theta/2$. This is called the Schottky anomaly, and can actually be observed in real systems. Isn't that nice?

We might ask ourselves at what sort of temperatures we might be able to observe the Schottky anomaly in the laboratory? It is useful to know that it is difficult to produce a permanent magnet with a field greater than about 1 Tesla. From our definition of θ, and as the magnitudes of μ_B and k_B are similar, you will see that in such a field the peak in the heat capacity will occur at a temperature of order 1 K!

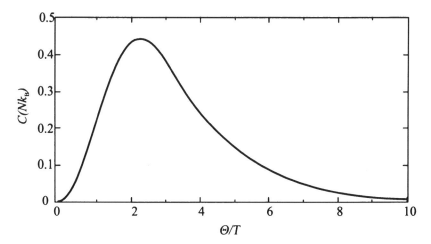

Fig. 3.3 Heat capacity of the spin-1/2 paramagnet as a function of temperature.

Having predicted the behaviour of U and C from the partition function, let us turn our attention to the entropy. Using eqns (2.33) and (2.31) we obtain

$$S = Nk_\mathrm{B} \ln Z_\mathrm{sp} + Nk_\mathrm{B} T \frac{\partial \ln Z_\mathrm{sp}}{\partial T}$$

$$= Nk_\mathrm{B} \ln \left[2 \cosh \left(\frac{\mu_\mathrm{B} B}{k_\mathrm{B} T} \right) \right] - \frac{N\mu_\mathrm{B} B}{T} \tanh \left(\frac{\mu_\mathrm{B} B}{k_\mathrm{B} T} \right) \ .$$

The limiting values of the entropy are

$$S \to Nk_\mathrm{B} \ln 2, \quad \text{when} \quad k_\mathrm{B} T \gg \mu_\mathrm{B} B \ ,$$

and

$$S \to 0, \quad \text{when} \quad k_\mathrm{B} T \ll \mu_\mathrm{B} B \ .$$

This is exactly what we would expect. At very low temperatures, all of the electrons must be in the lower state. There is only one way for all of them to be there (just as there is only one way of getting 10 heads when we flip a coin 10 times). So, $\Omega = 1$ and $S = k_\mathrm{B} \ln 1 = 0$. On the other hand, at extremely high temperatures $k_\mathrm{B} T$ is very large compared with the magnetic splitting between the levels, and in the limit of the temperature tending to infinity the magnetic splitting might as well not be there. In this case we are back to our normal coin tossing again – each state is equally likely – and we expect half to be in the upper level, and half in the lower level. We also know the number of ways this can happen, because when we worked it out for the coins it was $\Omega = 2^N$. Clearly, from $S = k_\mathrm{B} \ln \Omega$ this leads to an entropy of $Nk_\mathrm{B} \ln 2$.

Thus far we have derived U, C, S from the partition function. The last parameter we promised to derive was the magnetization. There is rather an elegant way of doing this, which will, we hope, make you realize the importance of the free energy, F, which we haven't yet used. From our knowledge of thermodynamics we know that for a magnetic system

$$dF = -MdB - SdT \quad , \tag{3.2}$$

and thus from eqns (2.34), (2.31) and (3.2) we find

$$M = -\left(\frac{\partial F}{\partial B}\right)_T = +Nk_\mathrm{B}T\left(\frac{\partial \ln Z_\mathrm{sp}}{\partial B}\right)_T \quad .$$

Convince yourself that this leads to

$$M = N\mu_\mathrm{B} \tanh\left(\frac{\mu_\mathrm{B} B}{k_\mathrm{B} T}\right) \quad .$$

In the limit of weak fields (i.e. $B \to 0$) this yields Curie's law of paramagnetism:

$$M = \frac{N\mu_\mathrm{B}^2 B}{k_\mathrm{B} T} \quad , \tag{3.3}$$

since $\tanh x \approx x$ for $x \ll 1$.

On the other hand at high magnetic fields the magnetization tends to a constant: $N\mu_\mathrm{B}$ (recall $\tanh x \to 1$ as $x \to \infty$). As we have discussed above, when the magnetic field is very high, the energy splitting between the levels will be large compared with $k_\mathrm{B}T$, and all the electrons will be in the ground state: that is to say, all of the magnetic moments of the electrons will be parallel to the magnetic field. Each electron has a magnetic moment μ_B, so if we have N of them pointing in the same direction the total magnetic moment saturates at a value of $N\mu_\mathrm{B}$.

Although you may not have thought of it in this way before, Curie's law is actually the equation of state of the magnetic system – it links the three variables (M, B, T) just as the equation of state of a gas in its usual form links (P, V, T). This is an important characteristic of the free energy F – it leads easily to the normal form of the equation of state of a system. Of course, we will see later on how the partition function of an ideal gas leads to its equation of state – we have not covered that so far because the form of the partition function for the gas is slightly more involved – but don't worry, we will get there soon!

3.2 Paramagnets with Angular Momentum J

The spin-1/2 paramagnet was a particularly simple system to consider. With only two energy levels the partition function was trivial to construct – indeed, this was one of the motivations for introducing it as our first use of the knowledge we have gained thus far. In general, however, a system can have more levels, and this will be the case for the paramagnet if the total angular momentum of the atom (orbital plus spin), J, is greater than 1/2. As we mentioned in Section 3.1, a level with angular momentum J in a magnetic field splits into $(2J+1)$ sublevels, with the z component of the angular momentum ranging from $-m_J$ to $+m_J$. For example, if $J = 2$, we will have 5 sublevels, with the z component of angular momentum taking on the values -2, -1, 0, 1, 2. In general, therefore, the partition function for a paramagnet will contain $(2J+1)$ terms in the summation.

Thus,

$$Z_{\rm sp} = \sum_{m_J=-J}^{+J} \exp(-g_J m_J \mu_B B/k_B T)$$

$$= \sum_{m_J=-J}^{+J} \exp(-Cm_J) \quad ,$$

where

$$C = \frac{g_J \mu_B B}{k_B T} \quad .$$

The partition function is thus a geometric progression with a finite number of terms:

$$Z_{\rm sp} = \frac{\exp(-CJ) - \exp(C(J+1))}{1 - \exp(C)}$$

$$= \frac{\exp(-C(J+1/2)) - \exp(C(J+1/2))}{\exp(-C/2) - \exp(C/2)}$$

$$= \frac{\sinh[(J+1/2)C]}{\sinh(C/2)} \quad .$$

Having derived the partition function, let us work out the magnetization:

$$M = Nk_B T \frac{\partial \ln Z_{\rm sp}}{\partial B} = g\mu_B \frac{\partial}{\partial C}\left[\ln\left(\frac{\sinh[(J+1/2)C]}{\sinh(C/2)}\right)\right]$$

$$= Ng\mu_B \left[(J+1/2)\coth[(J+1/2)C] - \frac{\coth(C/2)}{2}\right] \quad .$$

Recall that in the limit of small x, $\coth x \simeq 1/x + x/3$. Therefore, in the limit of small C

$$M = Ng_J\mu_B \left[\frac{(J+1/2)^2 C}{3} - \frac{C}{12}\right] = \frac{Ng_J^2\mu_B^2 J(J+1)B}{3k_B T} \quad . \qquad (3.4)$$

The term in square brackets is known as the Brillouin function, and eqn (3.4) gives us the low-field magnetization for the general paramagnet with angular momentum J, made up of orbital angular momentum L and spin

S.[2] Notice that this formula is consistent with the magnetization that we derived for the spin-1/2 paramagnet. That is to say, if we substitute $g_J = 2$ and $J = S = 1/2$ into eqn (3.4), it reduces to eqn (3.3).

In Fig. 3.4 we plot the experimentally measured magnetization for crystals containing Cr^{3+}, Fe^{3+} and Gd^{3+} ions. Notice that for small fields the magnetization is proportional to the magnetic field, as predicted by Curie's law. At high magnetic fields the magnetization saturates: once more, all of the electrons are in the lowest energy state, giving a total magnetization of $Ng_J\mu_B$. However, to observe such saturation experimentally, given that a strong magnetic field is of the order of 1 Tesla, we need to perform experiments at very low temperatures.

As for the spin-1/2 paramagnet, now that we have an expression for the partition function, we could also if we wished work out all the other thermodynamic quantities such as the internal energy, heat capacity, entropy etc.

3.3 The Simple Harmonic Oscillator

The paramagnet was a very simple system, insofar as it only had a finite number of equispaced energy levels and thus a finite number of terms in the partition function. We now turn our attention to a localized array (i.e. distinguishable particles) of simple harmonic oscillators (SHOs). Such an array of oscillators is perhaps the simplest model we could imagine for describing the vibrations of atoms in a crystal. In this simplified model (known as the Einstein model) the atoms in the crystal are assumed to vibrate as simple harmonic oscillators in all three dimensions (i.e. along the x, y, and z axes), as shown in Fig. 3.5. All the simple harmonic oscillators are taken to be identical, i.e. they all vibrate with the same frequency ν.

3.3.1 An Array of 1-D Simple Harmonic Oscillators

Before considering the full 3-dimensional model, for the moment for the sake of simplicity we will assume that our array of SHOs can only oscillate in 1 dimension (say along the z axis) – we will then extend the model to take into account all three dimensions. In contrast to the paramagnet there are an infinite number of energy levels, but the mathematics of calculating the partition function is still not too difficult, because the energy levels are all equispaced. The solution of Schrödinger's equation with the parabolic potential which defines an SHO leads to the energy spectrum:

$$E_n = (n + \frac{1}{2})h\nu$$

---

[2]The Landé g-factor for the general case is given by

$$g_J = \frac{3J(J+1) - L(L+1) + S(S+1)}{2J(J+1)} .$$

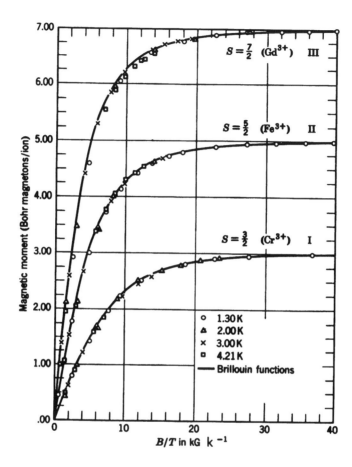

Fig. 3.4 Experimentally measured magnetization for crystals containing Cr^{3+}, Fe^{3+} and Gd^{3+} ions plotted as a function of B/T in units of kiloGauss per Kelvin (10^4 Gauss = 1 Tesla). (Reproduced from Fig. 3 of the classic paper by Warren E. Henry, published in *Phys. Rev.* **88**, 559 (1952).)

where ν is the frequency of the oscillator, h is Planck's constant, and here n stands for the nth energy level (not to be confused with the number of particles): n can take on any positive integer value including zero. It is trivial to show that the single-particle partition function of this 1-D SHO, $Z_{sp(1D)}$, is now a geometric progression:

$$Z_{sp(1D)} = \exp(-h\nu/2k_BT) \times \sum_{n=0}^{\infty} \exp(-nh\nu/k_BT)$$
$$= \frac{\exp(-h\nu/2k_BT)}{1 - \exp(-h\nu/k_BT)} \quad . \tag{3.5}$$

38 Paramagnets and Oscillators

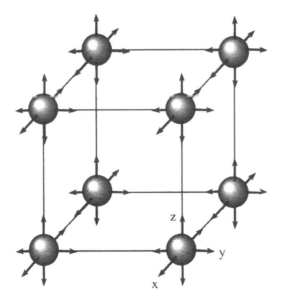

Fig. 3.5 In a very simple model of the vibrational motion of atoms in a crystal we consider the atoms to form an array of SHOs, each with natural frequency ν, that can vibrate in all three directions.

Again, now that we have found the partition function, in principle we know all of the thermodynamics. So to find the internal energy we substitute eqn (3.5) into eqn (2.25), and after some algebra we find

$$U = Nk_\mathrm{B}T^2 \frac{\partial \ln Z_{\mathrm{sp(1D)}}}{\partial T} = \frac{Nh\nu}{2} + \frac{Nh\nu}{\exp(h\nu/k_\mathrm{B}T) - 1} \quad . \tag{3.6}$$

Note the high-temperature limit:

$$U \to Nk_\mathrm{B}T \quad \text{as} \quad T \to \infty \quad .$$

We differentiate the expression for the internal energy, eqn (3.6), with respect to temperature to find the heat capacity:

$$C = \frac{\mathrm{d}U}{\mathrm{d}T} = Nk_\mathrm{B} \left(\frac{\theta}{T}\right)^2 \frac{\exp(\theta/T)}{(\exp(\theta/T) - 1)^2} \quad , \quad \text{where} \quad \theta = \frac{h\nu}{k_\mathrm{B}} \quad . \tag{3.7}$$

Note the similarity with the form of the heat capacity for the paramagnet (a '-' instead of a '+' in the denominator).

3.3.2 An Array of 3-D Simple Harmonic Oscillators

Until now we have allowed our SHOs to vibrate only in the z direction. However, clearly in a crystal the atoms can vibrate along any of the three

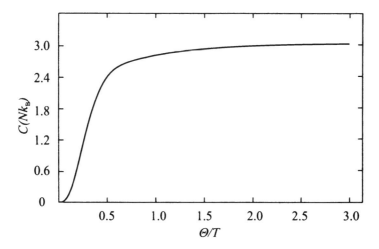

Fig. 3.6 Heat capacity of an array of 3-D simple harmonic oscillators. Temperature in units of θ – see text for definition.

orthogonal directions. Each of the three directions is independent, and thus the energy of the 3-D oscillator can be written as

$$E_{n_x,n_y,n_z} = (n_x + \frac{1}{2})h\nu + (n_y + \frac{1}{2})h\nu + (n_z + \frac{1}{2})h\nu \ .$$

As the three integers n_x, n_y, and n_z are independent (because the three directions are independent) the single-particle partition function of our 3-D SHO, $Z_{\text{sp(3D)}}$, can be written as

$$Z_{\text{sp(3D)}} = Z_{\text{sp}(x)} Z_{\text{sp}(y)} Z_{\text{sp}(z)} \ ,$$

where

$$Z_{\text{sp}(x)} = \exp(-h\nu/2k_\text{B}T) \times \sum_{n_x=0}^{\infty} \exp(-n_x h\nu/k_\text{B}T) \ ,$$

and similarly for y and z. Thus, comparing with eqn (3.5) we see that

$$Z_{\text{sp(3D)}} = (Z_{\text{sp(1D)}})^3 \ . \tag{3.8}$$

All of the thermodynamic paramaters we have discussed to date, such as U, S, and F, are functions of $\ln Z$. Therefore, given eqn (3.8), we immediately see that these parameters for an array of 3-D SHOs are simply three times the values of those for a 1-D SHO! Thus, we can immediately write down the heat capacity of the solid in this model: it is simply three times the value given in eqn (3.7):

$$C_{\text{3D}} = 3Nk_\text{B} \left(\frac{\theta}{T}\right)^2 \frac{\exp(\theta/T)}{(\exp(\theta/T) - 1)^2} \ , \tag{3.9}$$

where again $\theta = h\nu/k_\text{B}$.

This heat capacity is plotted as a function of temperature in Fig. 3.6. We find that at high temperatures ($T \gtrsim \theta$) the heat capacity is $3Nk_B$, which is the classical result. It turns out that for most materials, θ is of the order of room temperature, and if we substitute $T = \theta$ into eqn (3.9), we find that the heat capacity is $2.76Nk_B$, i.e. 92% of the maximum value. Thus, most solid materials at room temperature have roughly the classical value of the heat capacity, a result discovered by Dulong and Petit in the 19th century.

On the other hand, if we look at low temperatures such that $T \ll \theta$, we find from eqn (3.9) that the heat capacity tends to

$$C_{3D} \approx 3Nk_B \left(\frac{\theta}{T}\right)^2 \exp(-\theta/T) \ .$$

That is to say, the heat capacity falls off exponentially with temperature at low temperatures. This result was predicted by Einstein, and indeed is the main result of the Einstein model of heat capacities. This result was originally looked upon as a success, in that it explained the falloff at low temperatures. However, we will find out later (Section 8.4) that Einstein's model doesn't work(!) for a solid – the heat capacity actually falls off as T^3 at low temperatures, not exponentially. Einstein's assumption that all the oscillators are independent is not correct – in fact they are coupled together – and this makes the difference to the low-temperature form of the heat capacity. On the other hand, a 1-D SHO *is* a good description of the vibrational component of the heat capacity of a gas of diatomic molecules, as we shall discuss in Section 5.3.

3.4 Summary

- The single-particle partition function for a spin-1/2 paramagnet is:

$$Z_{sp} = 2\cosh\left(\frac{\mu_B B}{k_B T}\right) \ .$$

Using this we can work out the internal energy, heat capacity, entropy, magnetization, etc. For example, we found:

$$C = Nk_B \left(\frac{\theta}{T}\right)^2 \frac{\exp(\theta/T)}{(\exp(\theta/T) + 1)^2} \ , \quad \text{where} \quad \theta = \frac{2\mu_B B}{k_B}$$

and

$$M = N\mu_B \tanh\left(\frac{\mu_B B}{k_B T}\right) \ ,$$

which for small magnetic fields yields Curie's law:

$$M = \frac{N\mu_B^2 B}{k_B T} \ .$$

- For a paramagnet with angular momentum J and in small magnetic fields the magnetization is given by

$$M = \frac{N g_J^2 \mu_B^2 J(J+1) B}{3 k_B T},$$

which is consistent with the spin-1/2 result.
- For a localized array of 1-D simple harmonic oscillators

$$Z_{\text{sp(1D)}} = \frac{\exp(-h\nu/2k_B T)}{1 - \exp(-h\nu/k_B T)}.$$

Again we can use this to work out all of the thermodynamic parameters.
- The partition function for an array of SHOs that can vibrate in all three dimensions, $Z_{\text{sp(3D)}}$, is simply given by $(Z_{\text{sp(1D)}})^3$. As all the thermodynamic parameters are functions of $\ln Z_{\text{sp}}$, this means that the 3-D values are just 3 times the 1-D ones! For example, the heat capacity is

$$C_{\text{3D}} = 3 N k_B \left(\frac{\theta}{T}\right)^2 \frac{\exp(\theta/T)}{(\exp(\theta/T) - 1)^2}, \quad \text{where} \quad \theta = \frac{h\nu}{k_B}.$$

This is the Einstein heat capacity of a solid, which predicts a heat capacity of $3 N k_B$ at high temperatures and an exponential falloff at low temperatures.
- The most important point is that for any system all we need to do is to write down the partition function and then we know *all* of the thermodynamics.

3.5 Problems

1. A solid contains ions with a magnetic moment due to an angular momentum $J = 1$. Electric effects within the solid cause the states with $m_J = \pm 1$ to have an energy lower by Δ than the state with $m_J = 0$. A magnetic field B is applied parallel to the z axis so that the $m_J = \pm 1$ states shift in energy by $\pm g_J \mu_B B$.
(i) Show that the energies of the three states of the ion are 0, $-\Delta + g_J \mu_B B$, and $-\Delta - g_J \mu_B B$.
(ii) Construct the partition function and derive an expression for the susceptibility.
(iii) Curie's law states that the magnetic susceptibility, $\chi = \mu_0 M/B$, of a paramagnet (where μ_0 is the permeability of the vacuum) is inversely proportional to temperature. Show that Curie's law is obeyed (a) when $k_B T \gg \Delta$ and (b) when $k_B T \ll \Delta$, but not (c) when $k_B T \sim \Delta$.
(iv) Show that the ratio of the susceptibilities at a given temperature in cases (a) and (b) above is 3/2, and explain this result in elementary terms.

2. Given that the heat capacity of an array of N one-dimensional distinguishable simple harmonic oscillators is

$$C = \frac{dU}{dT} = Nk_B \left(\frac{\theta}{T}\right)^2 \frac{\exp(\theta/T)}{(\exp(\theta/T) - 1)^2}, \quad \text{where} \quad \theta = \frac{h\nu}{k_B},$$

find the high and low-temperature limits of C.

3. Show that the zero-point motion of an assembly of simple harmonic oscillators does not contribute to its entropy or heat capacity.

4. (i) Show that an assembly of N localized particles, each of which has a finite number, n, of energy levels with uniform spacing $\varepsilon = k_B\theta$, has a partition function Z^N, where

$$Z = \frac{Z(\theta/T)}{Z(n\theta/T)},$$

where, in this notation, $Z(\theta/T)$ is the partition function of a simple harmonic oscillator with characteristic temperature θ and $Z(n\theta/T)$ is the partition function of a simple harmonic oscillator with characteristic temperature $n\theta$. Hence, show that the heat capacity of this assembly can be written as the difference of two Einstein functions (i.e. as the difference in the heat capacities of two different simple harmonic oscillators).

(ii) Sketch the heat capacity as a function of temperature for (a) $n = 2$, (b) $n = 5$, and (c) $n \gg 1$. Comment on the results.

4
Indistinguishable Particles and Monatomic Ideal Gases

Physicists use the wave theory on Mondays, Wednesdays and Fridays, and the particle theory on Tuesdays, Thursdays and Saturdays.
W.H. Bragg

4.1 Distinguishable and Indistinguishable States

Up until this point we have been dealing with *distinguishable* particles. That is to say, we could tell which particle was which because they were *localized* within a solid. For instance, if one particle was in the ground state of a simple harmonic oscillator and a second was in the first excited state of another, we knew which was which (we could look and, for example, find the one in the ground state located at the corner of the crystal, and the excited one at some definite coordinate in the middle of the crystal). As noted, this comes about because the particles, although perhaps identical (e.g. they could both be carbon atoms in a diamond crystal), are localized within their own separate potentials.

However, consider what happens if one puts lots of identical particles in the same confining potential. This is what happens when we have a gas of identical particles – in this case all of the identical atoms are confined in the potential defined by the walls of the container (which can be approximated by an infinite square well potential). This changes everything, because now the states become *indistinguishable*. These ideas can be illustrated by reference to Fig. 4.1(a). Here we have drawn the wavefunctions of two particles in their separate potential wells in a solid: one is in the ground state, and one in an excited state. We swap the two particles. Can we tell the difference between the wavefunctions? The answer is obviously yes, the wavefunctions of the ground state and excited state have swapped positions in space – the particle in the ground state, which was on the left, is now on the right. On the other hand, in Fig. 4.1(b) we have drawn the wavefunctions for two identical gas particles confined in a box (i.e. an infinite square well, with which you are no doubt familiar from your studies in quantum mechanics). We now swap the particles. Can we tell the difference? No we cannot. Blindfold us, twirl us round, swap the particles, remove the blindfold, and we're none the wiser. We cannot tell the difference between the

44 *Indistinguishable Particles and Monatomic Ideal Gases*

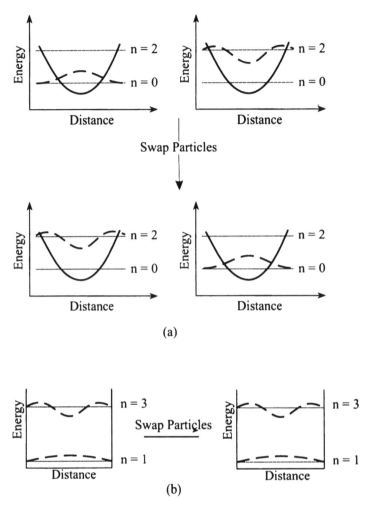

Fig. 4.1 Two identical potentials are confined in potential wells. The potentials are denoted by the solid lines, the magnitude of the wavefunctions by the dashed lines, and the energy levels by the dotted lines. If each of the two particles are in different wells, such as the quadratic potentials shown in (a), when we swap these localized (and thus distinguishable) particles, we can tell the difference. However, for identical particles in the same potential well (such as the infinite square well shown in (b)), we cannot tell the difference upon exchanging the particles.

two states. And importantly, as these are quantum, identical particles, we can never, ever tell the difference.

This phenomenon of indistinguishability is extremely important, and it totally changes the statistics. For a start, we find we cannot (unfortunately) use the simple method of 'number of microstates in a macrostate' that we used so successfully for the distinguishable particles. To prove this, look at

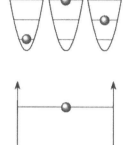

This is one of 6 microstates in the same macrostate.

As all the particles in the box are identical, there is only one way of arranging them, and our old counting method breaks down.

Fig. 4.2 For distinguishable particles localized in space, as shown in the upper part of the diagram, we can calculate the number of microstates in a macrostate, but for indistinguishable particles in the same potential well, as shown in the lower portion of the diagram the answer is 1 every time – our counting method breaks down.

Fig. 4.2, where we have represented a macrostate with one particle in each of the first three excited states. *If* the particles were distinguishable, there would be 3! = 6 ways of doing this. However, as the states are *indistinguishable* there is only 1 way (as we have just seen, now we can't tell if the particles are swapped). In fact, whatever macrostate we decide to look at, there only appears to be one microstate associated with it, so our method completely breaks down. Now, if we were taking the rigorous and sophisticated approach (which, certainly at this stage, we most definitely are not), this would be the point where we would introduce you to the delights of the method of ensembles developed by Gibbs.[1] However, we have promised to keep things as simple as possible, and so will use a different, less rigorous approach to solve this problem, which is set out in the next section.

4.2 Identical Gas Particles – Counting the States

It seems we are stuck. Our old method of counting the number of microstates in a given macrostate has let us down. Where do we go from here? As mentioned above, if we were to be pursuing the subject in a truly rigorous fashion, we would now need to resort to the method of ensembles developed by Gibbs. Luckily, in the case of a typical gas, there is simpler method we can adopt that relies on the fact that for such a gas the quantum energy levels are extremely closely spaced.

---

[1] For the stout of heart the method of ensembles is touched upon in Chapter 10; it acts as a bridge between the simpler approaches used in most of this book and more sophisticated treatments.

Consider a particle of mass m in a 3-D cubical box. We know from elementary quantum mechanics by solving Schrödinger's equation for a particle in a box that the energy of the level with principal quantum numbers n_x, n_y, and n_z is given by

$$\varepsilon_{n_x n_y n_z} = \frac{h^2}{8ma^2}(n_x^2 + n_y^2 + n_z^2) \ , \qquad (4.1)$$

where a is the length of the box and h is Planck's constant. Now typically for a gas at, say, room temperature, the integers n_x, n_y, and n_z are going to be large, because the energy of the particles puts them in very high quantum states. We can obviously rewrite eqn (4.1) as

$$n_x^2 + n_y^2 + n_z^2 = \frac{8ma^2 \varepsilon_{n_x n_y n_z}}{h^2} \ .$$

Therefore, the total number of states, $G(\varepsilon)$, with energies between 0 and ε is equal to the number of points (defined by the n_x, n_y, n_z) in the positive octant of a sphere of radius R, such that

$$R^2 = \frac{8ma^2 \varepsilon_{n_x n_y n_z}}{h^2} \ .$$

If the sphere is not too small (and we have already said the integers are going to be large numbers), then as the points have unit distance between them the number of points in the octant is given by the volume of the octant:

$$G(\varepsilon) = \frac{1}{8} \times \frac{4\pi}{3} \times \left(\frac{8ma^2 \varepsilon_{n_x n_y n_z}}{h^2}\right)^{3/2} = \frac{4\pi V}{3h^3}(2m\varepsilon)^{3/2} \ . \qquad (4.2)$$

What is this number, typically? Well, for nitrogen at standard temperature and pressure (STP), with a gram mole occupying 22.4 litres, the total number of available levels up to an energy of $3k_{\rm B}T/2$ is of order $10^{30}!$, a truly enormous number, even compared with the total number of molecules. Recall that this volume is occupied by 1 gram mole of the gas (Avogadro's number - 6×10^{23}). Thus we find that for a typical gas at typical temperatures we have a large number of exceedingly closely spaced, sparsely populated levels. This is the clue to how we solve the problem of the distribution function for these indistinguishable particles. What we do is to imagine collecting lots of levels together in bundles, with the ith bundle having g_i levels and n_i particles, as shown in Fig. 4.3. Then we make the (very good) approximation that all these g_i levels actually have the *same* energy, ε_i. For example, in the case of nitrogen described above we can divide the levels up into (say) 10^{10} different bundles, and the variation in energy in such a bundle will be about 1 part in 10^{10}. Each bundle will hold 10^{20} levels and just under 10^{14} particles.

The last point is extremely important: the number of levels is very much greater than the number of particles. And thus for a normal gas, it is very likely that each particle is in a different quantum state – and the probability that they are competing to get into the same state is so small that we can ignore it.

It may occur to you that under different conditions from STP, the above statement may not hold – and you would be correct. This leads to the fascinating topic of quantum statistics, which we will encounter in Chapter 6.

Now, to solve the problem of the distribution function, what do we want? Well, we need to find how many ways we can put n_i particles in such a bundle, regardless of energy (remember, we have made the approximation that all the g_i levels now have the same energy). For example, suppose we have $n = 3$ particles and $g = 10$ levels in a bundle (in practice, for a real gas, the ratio of states to particles is much higher than this), as shown in Fig. 4.4. We proceed by supposing the particles and the the partitions to be indistinguishable amongst themselves (i.e. we can't tell the difference betweeen each of the particles or each of the partitions, but we can tell the difference between a particle and a partition). Why do we do this? Well, even though the particles are indistinguishable (if we swap two we can't tell the difference), if we take the particle from one level and put it in another we now *can* tell the difference, because the energy level changes slightly in reality, even though we have called them degenerate for the purpose of solving the problem (think about it! –it's not necessarily obvious).

Firstly, if we could distinguish between each particle and each partition, then the first 'object' (particle or partition) could be picked in $(n + g - 1)$ ways and the second object in $(n + g - 2)$ ways, and so on – thus $(n + g - 1)!$ ways in total. But, as we stated above, if the particles and levels are indistinguishable amongst themselves we must divide by the total number

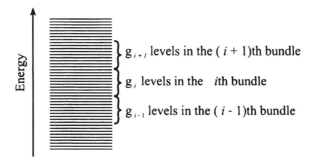

Fig. 4.3 For indistinguishable particles in a typical gas we group the energy levels into bundles and say the ith bundle is g_i degenerate.

48 Indistinguishable Particles and Monatomic Ideal Gases

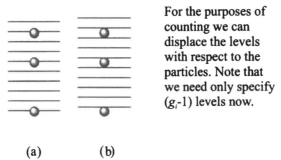

For the purposes of counting we can displace the levels with respect to the particles. Note that we need only specify $(g_i\text{-}1)$ levels now.

(a) (b)

Fig. 4.4 3 particles in a bundle with 10 levels. For the purposes of counting we displace the particles with respect to the levels to create partitions.

of ways of picking n particles and $(g - 1)$ partitions. So, finally we obtain

$$\text{Number of ways for one bundle} = \frac{(n_i + g_i - 1)!}{n_i!(g_i - 1)!} .$$

In our particular example this is 220 (check it!). So, taking into account all of the bundles, we obtain

$$\Omega = \prod_i \frac{(n_i + g_i - 1)!}{n_i!(g_i - 1)!} ,$$

and, taking into account that both n_i and g_i are very large, this can be approximated as

$$\Omega = \prod_i \frac{(n_i + g_i)!}{n_i!g_i!} . \tag{4.3}$$

Now, importantly for our gas at STP, we recall that $g_i \gg n_i$ and so the above equation can be simplified even further. We write

$$\Omega = \prod_i \frac{\overbrace{(n_i + g_i)(n_i + g_i - 1)\ldots(g_i + 1)}^{n_i \text{ terms}} (g_i)(g_i - 1)(g_i - 2)\ldots(2)(1)}{n_i!g_i!}$$

Because $g_i \gg n_i$, each of the n_i terms bracketed together in the above expression is approximately equal to g_i, and the expression simplifies to

$$\Omega = \prod_i \frac{g_i^{n_i}}{n_i!} . \tag{4.4}$$

The way forward from here should already be familiar to you from our treatment of the Boltzmann distribution for distinguishable particles back in Section 2.1 – once again we maximize Ω subject to the constraint that there is a fixed number of particles and also subject to the constraint that the energy is constant. Therefore, employing Stirling's theorem:

$$\ln \Omega = \sum_i [n_i \ln(g_i) - n_i \ln n_i + n_i] = \sum_i [n_i \ln\left(\frac{g_i}{n_i}\right)] + N \ .$$

Differentiating we find

$$\mathrm{d}\ln\Omega = \sum_i \ln\left(\frac{g_i}{n_i}\right) \mathrm{d}n_i \ ,$$

which must hold alongside

$$\sum_i \mathrm{d}n_i = 0 \ , \qquad \sum_i \varepsilon_i \mathrm{d}n_i = 0 \ .$$

We proceed as we did in Section 2.2, that is to say using the method of Lagrange multipliers. We leave it as an exercise for you to fill in the missing steps, and eventually find:

$$n_i = A g_i \exp(-\beta \varepsilon_i) \ ,$$

which is just the Boltzmann distribution again! Now β can be shown to be $1/k_B T$ as before, and A is again derived from the condition $N = \sum_i n_i$.

As there are many, many closely spaced levels, we often rewrite the above equation in terms of $n(\varepsilon)\mathrm{d}\varepsilon$, the number of particles in the energy interval between ε and $\varepsilon + \mathrm{d}\varepsilon$:

$$n(\varepsilon)\mathrm{d}\varepsilon = A g(\varepsilon) \exp(-\varepsilon/k_B T)\mathrm{d}\varepsilon \ , \tag{4.5}$$

where $g(\varepsilon)\mathrm{d}\varepsilon$ is a very important quantity known as the density of states. This tells us the number of quantum states with energies lying between ε and $\varepsilon + \mathrm{d}\varepsilon$. From eqn (4.2)

$$g(\varepsilon)\mathrm{d}\varepsilon = \frac{\mathrm{d}G(\varepsilon)}{\mathrm{d}\varepsilon}\mathrm{d}\varepsilon = \frac{4m\pi V}{h^3}(2m\varepsilon)^{1/2}\mathrm{d}\varepsilon \ . \tag{4.6}$$

Upon substituting eqn (4.6) into eqn (4.5) we obtain

$$n(\varepsilon)\mathrm{d}\varepsilon = A\frac{4m\pi V}{h^3}(2m\varepsilon)^{1/2}\exp(-\varepsilon/k_B T)\mathrm{d}\varepsilon \ . \tag{4.7}$$

We find A from the normalization condition:

$$N = \int_0^\infty n(\varepsilon)\mathrm{d}\varepsilon \ . \tag{4.8}$$

You can see from the above that by taking into account the indistinguishability of the identical particles we still obtain the Boltzmann distribution for the number of particles in a given energy range; so why did we bother? What is it that is different about the statistics in this case? It might, at first, appear as though nothing has changed from the distinguishable case. For example, we could once again derive the internal energy, U, in terms of the single-particle partition function. We will find we get the same result as for distinguishable particles. However, as we will show below, the *entropy S, is different* for indistinguishable particles. (Obviously F, the free energy, is also different, as $F = U - TS$).

To show that the entropy is different, let us work out all the thermodynamic quantities in exactly the same way as we did for distinguishable particles. Thus,

$$U = \sum_i A g_i \varepsilon_i \exp(-\varepsilon_i/k_B T) \quad ,$$

where A is given by

$$N = \sum_i A g_i \exp(-\varepsilon_i/k_B T) \quad .$$

Thus,

$$U = \frac{N \sum_i g_i \varepsilon_i \exp(-\varepsilon_i/k_B T)}{\sum_i g_i \exp(-\varepsilon_i/k_B T)}$$

$$= N k_B T^2 \frac{\partial}{\partial T} \left(\ln \sum_i g_i \exp(-\varepsilon_i/k_B T) \right)$$

$$= N k_B T^2 \frac{\partial}{\partial T} \ln Z_{sp} \quad .$$

Now for the entropy, $S = k_B \ln \Omega$, we find

$$S = k_B \left(\sum_i n_i \ln \left(\frac{g_i}{n_i} \right) + N \right)$$

$$= k_B \left(\sum_i n_i (-\ln A + \frac{\varepsilon_i}{k_B T}) + N \right)$$

$$= k_B (-\ln A + \frac{U}{k_B T} + N) \quad .$$

We find an expression for $\ln A$ by noting that A is obtained by normalization:

$$N = A \sum_i g_i \exp(-\varepsilon_i/k_B T) = A Z_{sp} \quad . \tag{4.9}$$

Thus,

$$\ln A = \ln N - \ln Z_{sp} \quad .$$

Hence,

$$S = k_B(-N \ln N + N \ln Z_{sp} + \frac{U}{k_B T} + N)$$

$$= k_B(N \ln Z_{sp} + \frac{U}{k_B T} - \ln N!)$$

$$= N k_B \ln Z_{sp} + N k_B T \frac{\partial \ln Z_{sp}}{\partial T} - k_B \ln N! \quad . \tag{4.10}$$

We now compare this equation for the entropy with that which we found for the *distinguishable* case – eqn (2.26). For distinguishable particles we found that

$$S = N k_B \ln Z_{sp} + N k_B T \frac{\partial \ln Z_{sp}}{\partial T} \quad .$$

So the entropy is different in the two cases – it is less in the indistinguishable case by $k_B \ln N!$ It should not surprise us that it is different and less: if the particles are indistinguishable, there must be less ways of arranging them, and thus there will be a lower entropy. It is interesting, however, to see that we can still write down the thermodynamic potentials in terms of the partition function of a whole system, Z_N, just as before – that is to say:

$$U = k_B T^2 \frac{\partial \ln Z_N}{\partial T} \quad ,$$

$$S = k_B \ln Z_N + k_B T \frac{\partial \ln Z_N}{\partial T} \quad , \tag{4.11}$$

$$F = -k_B T \ln Z_N \quad . \tag{4.12}$$

But now this only works if we define

$$Z_N = \frac{Z_{sp}^N}{N!} \tag{4.13}$$

(remember for *distinguishable* particles $Z_N = Z_{sp}^N$). Again, the difference is due to this feature of indistinguishability, which reduces the number of ways of arranging the particles.

4.3 The Partition Function of a Monatomic Ideal Gas

Having worked out the thermodynamic functions in terms of the partition function of the system (or single-particle partition function), we are now in a position where we are able to work out that partition function for an ideal gas. Although up to now we have written the partition function as a series, in this case we have so many closely spaced levels that we are justified in writing it as an integral:

$$Z_{\text{sp}} = \int_0^\infty g(\varepsilon) \exp(-\varepsilon/k_B T) d\varepsilon$$
$$= \int_0^\infty \frac{4m\pi V}{h^3} (2m\varepsilon)^{1/2} \exp(-\varepsilon/k_B T) d\varepsilon \quad .$$

This can be integrated (see Appendix B) to give the result:

$$Z_{\text{sp}} = V \left(\frac{2\pi m k_B T}{h^2} \right)^{3/2} \quad . \tag{4.14}$$

It is important to realize that this is the single-particle partition function related solely to the translational energy of the gas particles. When we worked out the energy levels (which gave us $g(\varepsilon)d\varepsilon$), we were only talking about the translational energy of the particles confined in the box. If the gaseous atoms are diatomic (or greater), then they can also have rotational and vibrational energy. We will deal with those other degrees of freedom in Section 5.1.

4.4 Properties of the Monatomic Ideal Gas

So we have derived the single-particle partition function – eqn (4.14). Let us now see what this tells us about the properties of the gas. First, we start with the internal energy:

$$U = k_B T^2 \frac{\partial \ln Z_N}{\partial T} = k_B T^2 \frac{\partial (N \ln Z_{\text{sp}} - \ln N!)}{\partial T} \quad . \tag{4.15}$$

Substituting eqn (4.14) into eqn (4.15) we obtain

$$U = N k_B T^2 \frac{\partial (\ln T^{3/2} + \ln(\ldots))}{\partial T} = \frac{3 N k_B T}{2} \quad ,$$

where the $\ln(\ldots)$ denotes all the other terms and constants in the partition function that do not depend on T, and therefore these will drop out when we differentiate with respect to T. So we obtain the internal energy of the gas consistent with the standard results of kinetic theory, and the heat capacity due to the translational energy is obviously $3Nk_B/2 = 3R/2$ per mole. We have stressed previously that all of the thermodynamics is present within the partition function, and thus we should be able to derive the equation of state of the gas from it. How do we do this? Well, from thermodynamics we know that

$$dF = -SdT - PdV \quad ,$$

and thus

$$P = -\left(\frac{\partial F}{\partial V}\right)_T = k_B T \left(\frac{\partial \ln Z_N}{\partial V}\right)_T . \quad (4.16)$$

where we have made use of eqn (4.12). Note that the expression for the pressure in eqn (4.16) contains a partial derivative with respect to V of the logarithm of the partition function, and from eqns (4.14) and (4.13) we see

$$\ln Z_N = N \ln V + \ln(\text{other terms}) .$$

These other terms will be zero on differentiation with respect to V. Thus,

$$P = k_B T \left(\frac{\partial [N \ln V]}{\partial V}\right)_T = \frac{N k_B T}{V} ,$$

as expected. Perhaps you can now start to appreciate the usefulness of the partition function. Just to prove how powerful it is, we will now show how we can derive the formula for the adiabat of the gas. First, we note that the expression for the entropy of the gas can be found by substituting eqn(4.14) into eqn (4.10). After some algebra we find

$$S = N k_B \left[\ln V + \frac{3}{2}\ln T + \frac{3}{2}\ln\left(\frac{2\pi m k_B}{h^2}\right) - \ln N + \frac{5}{2}\right] . \quad (4.17)$$

This is known in the trade as the Sackur–Tetrode equation.[2] Consider an isentropic ($\Delta S = 0$) expansion of the gas from (V_1, T_1) to (V_2, T_2). As so many terms remain constant, it is trivial to show that

$$\ln V_1 + \frac{3}{2}\ln T_1 = \ln V_2 + \frac{3}{2}\ln T_2 , \quad \text{i.e.} \quad VT^{3/2} = \text{constant} .$$

Combined with the equation of state, this yields $PV^{5/3}$ = constant, the formula for an adiabat. Once again we see that *all* of the thermodynamics is contained within the partition function.

4.5 More about Adiabatic Expansions

We have found that for a monatomic ideal gas during an adiabatic expansion, $VT^{3/2}$ is constant. We did this by substituting the partition function for the gas into our formula for the entropy. However, there is another way of deriving this result which gives additional insight into the pertinent physics involved.

---

[2] Note that as $\ln V - \ln N = \ln V/N$ everything in the square brackets of eqn (4.17) remains the same if we double the number of particles and the volume. This important (and correct) result would not have been obtained had we erroneously put $Z_N = (Z_{sp})^N$.

Recall that we convinced ourselves (see eqn (2.15)) that the statistical term corresponding to changes in entropy was $TdS = \sum_i \varepsilon_i dn_i$, and thus along an isentrope

$$\frac{1}{T}\sum_i \varepsilon_i dn_i = 0 \quad . \tag{4.18}$$

It is of interest to consider the implications of this equation for the adiabatic expansion of an ideal monatomic gas. Remember that we treat the gas container as a 3-D infinite square well with sides of length a – the energy levels being given by the solutions to Schrödinger's equation, that is to say

$$\varepsilon_{n_x n_y n_z} = \frac{h^2}{8ma^2}(n_x^2 + n_y^2 + n_z^2) \quad .$$

An expansion corresponds to the walls of the container getter further apart (i.e. a increasing). Clearly the quantum energy levels change during such an expansion – indeed each level will fall as a^{-2}. Note that all of the quantum levels fall as the same function of a. Therefore, in order for eqn (4.18) to be satisfied, the actual number of particles in a particular level does not change during the expansion. This is quite a neat picture – we can think of all of the energy levels, with their various populations which fall off exponentially. We let the gas expand adiabatically, and all that happens is that the levels get closer together – the populations of each level remain constant, as shown in Fig. 4.5. But we also know that Boltzmann's law must be obeyed, i.e.

$$n_i = Ag_i \exp(-\varepsilon_i/k_B T) \quad .$$

We have argued that during the adiabatic expansion the number of particles in a given level must remain constant. For this to be true then it is evident that ε_i/T must remain constant to satisfy the Boltzmann distribution. Following this line of argument still further, given the form of the energies of the box, we conclude that along an adiabat for a monatomic gas

$$a^2 T = \text{constant} \quad .$$

But the volume of the box is given by $V = a^3$, and thus

$$V^{2/3} T = \text{constant} \quad ,$$

which is just the formula for the adiabat for an ideal monatomic gas once more! We deduced this simply from the concept that the populations of the levels could not change during an adiabatic expansion of a monatomic gas.

There are analogous arguments we can put forward for the case of a paramagnet during adiabatic demagnetization (i.e. reducing the B field

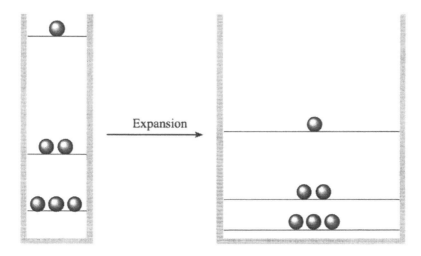

Fig. 4.5 When a monatomic ideal gas expands adiabatically, the populations in each level remain the same.

adiabatically). Once more, as above, we argue that ε_i/T must be constant during an adiabatic change. However, now the energy levels of a paramagnet depend linearly (at least for weak fields) on the magnetic field – for example recall that a simple spin-1/2 paramagnet has energy levels $\pm\mu_B B$ in the presence of a magnetic field. Therefore, we conclude that the parameter B/T must be constant during adiabatic demagnetization.

Simple arguments such as these can be very powerful, enabling us to deduce the form of the adiabat without even resorting to determining the partition function. However, caution should be exercised: this simple method works for finding the form of the adiabat because all of the levels have their energy scaled in the same way (i.e. for the monatomic gas they all scale as $V^{-2/3}$, and for the magnet all the energy levels are proportional to B). The situation is not always so simple. For example, in the next chapter we shall consider the case of a gas of diatomic molecules that can rotate as well as have translational motion. In this case the spacing of the rotational energy levels does not alter when the box expands, but the translational energy levels do change. In this case we cannot argue that the populations of individual levels remain the same during the adiabatic expansion, and the situation becomes slightly more complicated. That said, we have, in fact, already found in Section 4.4 what actually remains constant in the general case – it is the partition function, $Z_{\rm sp}$. This is also consistent with our argument about the monatomic gas and the paramagnet. Recall that the partition function of the monatomic gas is

$$Z_{\rm sp} = V \left(\frac{2\pi m k_{\rm B} T}{h^2}\right)^{3/2}.$$

56 Indistinguishable Particles and Monatomic Ideal Gases

If we say that the partition function is constant during an adiabatic expansion, then this is equivalent to saying $VT^{3/2}$ is constant. For the paramagnet, the partition function is a function of B/T, and thus this is constant. So what we have really learnt is the extra insight that in the particular case of an adiabatic change where the energy levels all scale in the same way, the energy levels change but their populations do not.

4.6 Maxwell–Boltzmann Distribution of Speeds

As a short aside it is worth noting that we have already effectively derived the Maxwell-Boltzmann distribution of speeds. This is because in Section 4.7 we derived the number of particles per unit energy interval, i.e.

$$n(\varepsilon)d\varepsilon = Ag(\varepsilon)\exp(-\varepsilon/k_BT)d\varepsilon$$
$$= A\frac{4\pi mV}{h^3}(2m\varepsilon)^{1/2}\exp(-\varepsilon/k_BT)d\varepsilon \quad .$$

Now, A can be deduced from normalization, and we found previously in eqn (4.9) that

$$A = N/Z_{\text{sp}} = \frac{N}{V}\left(\frac{h^2}{2\pi mk_BT}\right)^{3/2} \quad ,$$

leading to

$$n(\varepsilon)d\varepsilon = \frac{2\pi N}{(\pi k_BT)^{3/2}}(\varepsilon)^{1/2}\exp(-\varepsilon/k_BT)d\varepsilon \quad .$$

To get to the Maxwell–Boltzmann distribution of speeds, we simply note that the kinetic energy of a particle is given by

$$\varepsilon = \frac{mc^2}{2} \quad ,$$

yielding

$$n(c)dc = N\left(\frac{2}{\pi}\right)^{1/2}\left(\frac{m}{k_BT}\right)^{3/2}\exp(-mc^2/2k_BT)c^2dc \quad .$$

This distribution is plotted in Fig. 4.6, and typical velocities of gases at STP are discussed in problem 3 in Section 4.9.

4.7 Gibbs Paradox

The Gibbs paradox was a classical problem related to distinguishable and indistinguishable particles. Consider Fig. 4.7: there are two gases, A and B, which each contain N particles of masses m_A and m_B respectively, and each occupies a volume V. They are separated by a partition, as shown.

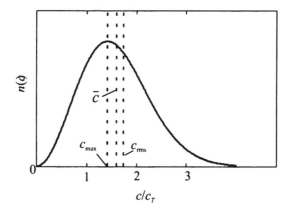

Fig. 4.6 The distribution of speeds in a Maxwell–Boltzmann gas. c_T is defined as $\sqrt{k_B T/m}$. The points of the most probable speed, $c_{\max}$, the mean speed, $\bar{c}$, and the root mean square speed, $\sqrt{\overline{c^2}}$, are shown.

The partition is removed and the gases are allowed to mix. The question is what happens to the entropy if the two gases are (i) different and (ii) the same? The paradox arose because this concept of indistinguishability was not initially understood.

The issue is as follows. Before the removal of the partition we have two gases, each with N particles, each separately occupying volumes V. The entropy is obviously the sum of the two separate entropies. So, the total initial entropy is

$$S_{\text{Initial}} = Nk_B \left[\ln V + \frac{3}{2}\ln T + \frac{3}{2}\ln\left(\frac{2\pi m_A k_B}{h^2}\right) - \ln N + \frac{5}{2}\right] +$$
$$Nk_B \left[\ln V + \frac{3}{2}\ln T + \frac{3}{2}\ln\left(\frac{2\pi m_B k_B}{h^2}\right) - \ln N + \frac{5}{2}\right] \quad . \quad (4.19)$$

Now the partition is removed, and the gases allowed to mix. Intuitively, we know that the entropy must increase, as we have introduced a further degree of disorder into the system – but how do we prove this, and what is the increase in entropy? The salient point is that although the particles in each gas are indistinguishable amongst themselves, we can tell the difference *between* the two gases – one could be helium, say, and the other neon. When the barrier is removed and they are allowed to mix, we now have two gases, each with N particles, occupying a volume $2V$. The partition functions of each gas are different (as they have different masses), and the partition function of the total mixed system is now

$$Z_{\text{system}} = Z_{\text{gasA}} \times Z_{\text{gasB}} \quad ,$$

as the gas A is distinguishable from the gas B. However, as the atoms in each gas are indistinguishable amongst *themselves*

58 Indistinguishable Particles and Monatomic Ideal Gases

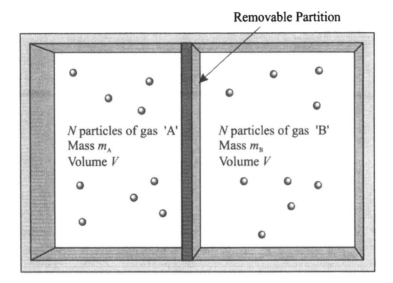

Fig. 4.7 Two gases are separated by a removable partition. The Gibbs paradox is the question of what happens to the entropy when the partition is removed if the gases are (i) different and (ii) the same.

$$Z_{\text{gasA}} = \frac{Z_{\text{spA}}^N}{N!} \quad , \text{ and } \quad Z_{\text{gasB}} = \frac{Z_{\text{spB}}^N}{N!} \quad .$$

Therefore,

$$Z_{\text{system}} = \frac{Z_{\text{spA}}^N Z_{\text{spB}}^N}{N!N!} \quad . \tag{4.20}$$

where each single-particle partition function, Z_{spA} and Z_{spB}, is proportional now to $2V$. The form of eqn (4.20) is important – compare it with the relationship between Z_N and Z_{sp} for distinguishable particles (eqn (2.31)), and make sure you understand why the denominator is $N!N!$. When eqn (4.20) is substituted into the formula for the entropy in terms of the partition function of a system (eqn (4.11)) we find

$$S_{\text{Final}} = Nk_B \left[\ln 2V + \frac{3}{2}\ln T + \frac{3}{2}\ln\left(\frac{2\pi m_A k_B}{h^2}\right) - \ln N + \frac{5}{2}\right] +$$

$$Nk_B \left[\ln 2V + \frac{3}{2}\ln T + \frac{3}{2}\ln\left(\frac{2\pi m_B k_B}{h^2}\right) - \ln N + \frac{5}{2}\right]$$

$$= S_{\text{Initial}} + 2Nk_B \ln 2 \quad .$$

So the entropy of each of the gases has increased by $Nk_B \ln 2$ due to the mixing process. This tells us that it is exceedingly unlikely that the two gases will 'unmix' spontaneously in any sensible time.

However, consider the effect of the removal of the partition for the case when the two gases are exactly the same. Surely the entropy cannot increase then, because after a while we could reinsert the partition, and would have two separate gases again, but not be able to tell the difference from the initial set-up. How does this fit in with the above argument that showed that there was an increase in entropy due to mixing? (This was the paradox.) [3]

Intuitively we know the entropy cannot increase for these indistinguishable particles, but how do we show this mathematically? Well, the point is that after the partition is removed, it looks as though we have $2N$ *indistinguishable* particles in a total volume of $2V$, so in terms of the single-particle partition function, Z_{sp} (where the volume in Z_{sp} is now again $2V$) the partition function of the system of the two gases together is now

$$Z_{\text{system}} = \frac{Z_{sp}^{2N}}{(2N)!} \quad . \tag{4.21}$$

Note that eqn (4.21) is different from eqn (4.20) just as Z_N is related to Z_{sp} in a different way for distinguishable and indistinguishable particles. When eqn (4.21) is substituted into the formula for the entropy, i.e. eqn (4.11), this gives us the same entropy as eqn (4.19) – try it! So for the *same* gases the entropy does not change when we remove the barrier.

In the distinguishable case, how different do the gases need to be to get the entropy change of $Nk_B \ln 2$ per gas? Well, as close as we like - as long as we can somehow tell the difference. For example, they could be completely different (helium and neon), or they could be two different isotopes of the same element, it doesn't matter. As long as there is *some* way of telling the difference, we will increase the entropy upon mixing.

4.8 Summary

- The particles in a gas, unlike those localized in a solid, are *indistinguishable*.
- To solve the statistics we group particles into bundles that are g_i degenerate. For a typical gas at STP $g_i \gg n_i$, and so we assume there are no states with more than one particle. The number of arrangements then becomes

$$\Omega = \prod_i \frac{(g_i^{n_i})}{n_i!} \quad .$$

---

[3] Indeed the argument is so serious that classical macroscopic thermodynamics fails at this point since it does not recognize the concepts of distinguishability and indistinguishability!

- This leads to the Boltzmann distribution, as for distinguishable particles:
$$n_i = A g_i \exp(-\beta \varepsilon_i) \quad ,$$
with $\beta = 1/k_B T$.
- As the levels are closely spaced, we can replace the degeneracy g_i by a continuous quantity $g(\varepsilon)\mathrm{d}\varepsilon$, which is the number of states with energies between ε and $\varepsilon+\mathrm{d}\varepsilon$ and is known as the density of states. The number of particles with energies within this same range, $n(\varepsilon)\mathrm{d}\varepsilon$, is then given by
$$n(\varepsilon)\mathrm{d}\varepsilon = A g(\varepsilon) \exp(-\varepsilon/k_B T)\mathrm{d}\varepsilon \quad ,$$
where A is now defined by the condition
$$N = \int_0^\infty n(\varepsilon)\mathrm{d}\varepsilon \quad .$$

- The single-particle partition function of an ideal gas is
$$Z_{\mathrm{sp}} = V \left(\frac{2\pi m k_B T}{h^2}\right)^{3/2} \quad .$$

- In terms of the *single-particle* partition function, Z_{sp} the entropy of indistinguishable particles is
$$S = N k_B \ln Z_{\mathrm{sp}} + N k_B T \frac{\partial \ln Z_{\mathrm{sp}}}{\partial T} - k_B \ln N! \quad ,$$
which is $k_B \ln N!$ less than for the distinguishable case. Note, however, that the entropy has the same form as a function of the partition function of the *system*, i.e.
$$S = k_B \ln Z_N + k_B T \frac{\partial \ln Z_N}{\partial T} \quad ,$$
as long as we now define
$$Z_N = \frac{Z_{\mathrm{sp}}^N}{N!}$$
(as opposed to the distinguishable case, where $Z_N = Z_{\mathrm{sp}}^N$).
- From the single-particle partition function we can derive the heat capacity of the gas, the equation of state, the formula for an adiabatic expansion – in fact we know all the thermodynamic properties of the system.
- Upon mixing two gases, both initially containing N particles in a volume V, there is an increase in entropy of $N k_B \ln 2$ for each gas if we can distinguish between the two gases. If the two gases cannot be told apart, there is no change in entropy upon mixing.

4.9 Problems

1. Show that the susceptibility, χ, of N spin-1/2 paramagnetic atoms in a volume V does not depend on whether or not they are localized (i.e. is the same for distinguishable and indistinguishable statistics).
2. Consider a gas where the particles can only move on the surface of a plane and thus are restricted to two-dimensional motion. Using a similar argument to that put forward in Section 4.2, find the density of states, partition function, and internal energy for this two-dimensional system.
3. What is the root-mean-square speed of a nitrogen molecule at room temperature?
4. Show that the ratio of the mean-square-speed to the square of the mean speed of the particles in a gas is $3\pi/8$.

5
Diatomic Ideal Gases

We shall never get people whose time is money to take much interest in atoms.
S. Butler

5.1 Other Degrees of Freedom

Polyatomic molecules can rotate and vibrate, as well as have translational motion. The number of degrees of freedom a molecule possesses depends on the number of atoms within it. Here we consider the simplest case – diatomic molecules. The single-particle partition function that we have dealt with thus far has only taken into account the translational motion of the gas, and has not included these rotational and vibrational effects.

We must now figure out how to treat the rotational and vibrational parts of the problem. The way we do this is to assume that they are separate, in the sense that there is no link between the translational, rotational or vibrational energy levels the molecule occupies. This implies that the total energy levels present within the system are simply the sums of the individual energy components:

$$\varepsilon_{\text{total}} = \varepsilon_{\text{trans}} + \varepsilon_{\text{rot}} + \varepsilon_{\text{vib}}$$

and, importantly, that the probability of being in a particular overall energy level, $\varepsilon_{\text{total}}$, can be found by multiplying together the probabilities of being in a particular translational, rotational, and vibrational level:

$$g_{\text{total}} \exp(-\varepsilon_{\text{total}}/k_{\text{B}}T) = $$
$$g_{\text{trans}} \exp(-\varepsilon_{\text{trans}}/k_{\text{B}}T) \times g_{\text{rot}} \exp(-\varepsilon_{\text{rot}}/k_{\text{B}}T) \times g_{\text{vib}} \exp(-\varepsilon_{\text{vib}}/k_{\text{B}}T) \quad .$$

This in turn implies that the resultant single-particle partition function is of the form

$$Z_{\text{sp}} = Z_{\text{trans}} \times Z_{\text{rot}} \times Z_{\text{vib}} = V \left(\frac{2\pi m k_{\text{B}} T}{h^2} \right)^{3/2} \times Z_{\text{rot}} \times Z_{\text{vib}} \quad .$$

Notice that this form makes sense in that the total energy of the gas, U, depends on $\ln Z$, and therefore we find $U = U_{\text{trans}} + U_{\text{rot}} + U_{\text{vib}}$ as we would expect. We now need to determine Z_{rot} and Z_{vib}.

5.2 Rotational Heat Capacities for Diatomic Gases

From quantum mechanics we find that for a rigid rotator the rotational energy levels are given by

$$\varepsilon_J = \frac{\hbar^2}{2I} J(J+1) \quad , \qquad (5.1)$$

where J is an integer, and I is the moment of inertia of the molecule.[1]

We also know from quantum mechanical considerations that the degeneracy of a particular energy level associated with angular momentum J is $(2J+1)$ (i.e. all the different values of m_J – the z component of the angular momentum – run from $-J$ to $+J$ in steps of 1), and so the rotational partition function is [2]

$$Z_{\rm rot} = \sum_{J=0}^{\infty} (2J+1) \exp\left[-\hbar^2 J(J+1)/2Ik_{\rm B}T\right] \quad . \qquad (5.2)$$

At this juncture it is interesting to work out the approximate spacing between energy levels and compare it with, say, room temperature. Consider the difference in energy, $\Delta\varepsilon$, between the $J=0$ and $J=1$ rotational energy levels. From eqn (5.1):

$$\Delta\varepsilon = \frac{\hbar^2}{I} \quad . \qquad (5.3)$$

We now define by convention

$$\theta_{\rm rot} = \frac{\hbar^2}{2Ik_{\rm B}} \quad .$$

This 'rotational temperature' represents the typical temperature necessary to have a reasonable fraction of the molecules in the first rotational state. At temperatures much below this, the molecule will hardly be rotating at all – almost all of the molecules will be in the rotational ground state. How high is this temperature in practice for typical diatomic molecules, and how

---

[1] Classically the energy of a rigid rotator is $L^2/2I$, where L is the angular momentum. To obtain the quantum mechanical solution we use the angular momentum operator, $\hat{L}$, and the Hamiltonian becomes

$$\frac{\hat{L}^2}{2I}\phi = \varepsilon\phi \quad ,$$

where ϕ is the wavefunction. The eigenvalues of the angular momentum operator are $\hbar^2 J(J+1)$, where J is an integer, and thus we obtain eqn (5.1).

[2] The treatment of the rotational part of the partition function that we give here is somewhat incomplete. In practice one has to take into account the nuclear spin of the two atoms in the molecule – this is especially important for hydrogen at low and intermediate temperatures. However, at high temperatures all diatomic molecules exhibit a heat capacity of R per mole.

does it compare with room temperature? To answer this question we need to know the moment of inertia, I – that is to say, the mass of the atoms in the molecule and the distance separating them. Take, for example, the nitrogen molecule. It comprises two nitrogen atoms, of atomic mass 14, separated by a distance of 0.11 nm. From this we calculate the moment of inertia, $\sum mr^2$, and substituting into eqn (5.3) find $\theta_{\text{rot}} = 2.8$ K. A similar calculation for hydrogen gives $\theta_{\text{rot}} = 82$ K. These values should be compared with room temperature, which is nearly 300 K. Thus their moment of inertia is such that room temperature is very large with respect to the separation of the lower energy levels for all diatomic molecules, except perhaps for hydrogen.

Therefore, in most gases the rotational levels are closely spaced compared with $k_B T$ at room temperature, and under these circumstances we are justified in replacing the summation by an integral. Let

$$x^2 = J(J+1) \ .$$

Then the rotational partition function can be written

$$Z_{\text{rot}} = \int_0^\infty 2x \exp(-\alpha x^2) dx \ ,$$

where

$$\alpha = \frac{\hbar^2}{2Ik_B T} = \frac{\theta_{\text{rot}}}{T} \ .$$

Thus,

$$Z_{\text{rot}} = \frac{1}{\alpha} = \frac{2Ik_B T}{\hbar^2} = \frac{T}{\theta_{\text{rot}}} \ .$$

This is the high-temperature approximation for the rotational part of the partition function of a diatomic gas. Using this high-temperature form (i.e. $T \gg \theta_{\text{rot}}$) of the partition function, we can work out the internal energy of the molecule due to rotations at such temperatures:

$$U = Nk_B T^2 \frac{\partial \ln(T/\theta_{\text{rot}})}{\partial T} \to Nk_B T \quad \text{at high} \ T \ ,$$

and

$$C_V(\text{rot}) = Nk_B = R \quad \text{per mole} \ .$$

As we emphasized above, room temperature counts as being high as far as the rotational levels are concerned, so a heat capacity of R due to rotations is a good approximation for most diatomic molecules. We notice that even for hydrogen, which has the highest value of θ_{rot}, room temperature is about 3.5 θ_{rot}. Clearly if we consider the gas at low temperatures, the situation will be different. As the temperature is lowered, fewer rotational levels will be occupied, and eventually, at very low temperatures where $T < \theta_{\text{rot}}$, only the ground state and perhaps the first rotational level will be significantly occupied. Evidently, at such a point we cannot use an integral to evaluate the partition function. Indeed, at such low temperatures (i.e. when $\alpha \gg 1$) the partition function can be well approximated by the first two terms in the series given by eqn (5.2)

$$Z_{\text{rot}} = 1 + 3\exp(-2\theta_{\text{rot}}/T) \quad ,$$

i.e. we include just the ground and first excited state. Note that although the second term is small compared with 1, we need to keep it if we are to find the temperature dependence of any of the thermodynamic functions (e.g. to work out the heat capacity) because otherwise there is nowhere for the energy to go as the system warms up. We leave it as an exercise for you to show that the above equation implies that the heat capacity at low temperatures falls off exponentially.

5.3 The Vibrational Partition Function of a Diatomic Gas

Because of the bond between the two atoms, the atoms in the diatomic molecule can vibrate. To a reasonable approximation, we can treat this behaviour as a 1-D simple harmonic oscillator, which we have already covered in Section 3.3. Recall that for a 1-D SHO the nth energy level is given by

$$E_n = (n + \frac{1}{2})h\nu \quad ,$$

where ν is the frequency of the oscillator. Thus, the single-particle partition function is given by

$$Z_{\text{vib}} = \exp(-h\nu/2k_\text{B}T) \sum_{n=0}^{\infty} \exp(-nh\nu/k_\text{B}T)$$
$$= \frac{\exp(-h\nu/2k_\text{B}T)}{1 - \exp(-h\nu/k_\text{B}T)} \quad .$$

Using this partition function, we found the heat capacity to be

$$C = \frac{dU}{dT} = Nk_\text{B} \left(\frac{\theta_{\text{vib}}}{T}\right)^2 \frac{\exp(\theta_{\text{vib}}/T)}{(\exp(\theta_{\text{vib}}/T) - 1)^2}$$

where now the characteristic temperature is given by

$$\theta_{\text{vib}} = \frac{h\nu}{k_\text{B}}$$

This heat capacity decays exponentially as $T \ll \theta_{\text{vib}}$, and tends to Nk_B (i.e. R per mole) as $T \gg \theta_{\text{vib}}$. Importantly, and in contrast with the rotational

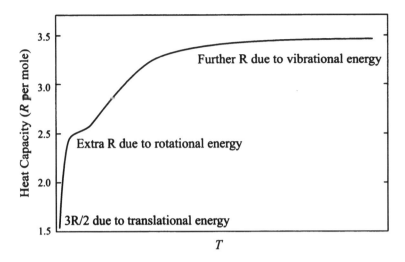

Fig. 5.1 Schematic diagram of the heat capacity of a diatomic molecule. The heat capacity per mole is 3R/2 due to translational energy, rapidly rising to 5R/2 at low temperatures due to rotation, and increasing to 7R/2 at high temperature due to vibrations. For a typical diatomic molecule, room temperature is such that the molecule is fully rotating, but not vibrating, and it has a heat capacity of 5R/2 at such a temperature.

temperatures, these vibrational temperatures tend to be large compared with room temperature. For example, for H_2, N_2, and Cl_2 the value of θ_{vib} is 6210 K, 3340 K, and 810 K respectively. Note how the vibrational temperature drops for heavier molecules. Thus we need to heat diatomic gases to quite high temperatures before we start to see the effects of the vibrational energy levels. Diatomic gases at room temperature therefore have heat capacities very close to $5R/2$ (with $3R/2$ due to translational motion, and R due to rotations), only rising to $7R/2$ at much higher temperatures. The heat capacity of a diatomic gas is plotted in schematic form as a function of temperature in Fig. 5.1. It was the experimental observation of this temperature dependence of the heat capacity that contributed to the acceptance of quantum theory.

5.4 Putting it All Together for an Ideal Gas

Thus we see that for an ideal gas

$$Z_{\text{sp}} = Z_{\text{trans}} \times Z_{\text{rot}} \times Z_{\text{vib}} = V \left(\frac{2\pi m k_B T}{h^2} \right)^{3/2} \times Z_{\text{rot}} \times Z_{\text{vib}} \ .$$

There are a couple of important points to note about the above equation. First, recall that neither Z_{rot} and Z_{vib} depend on V, and so $Z_{\text{sp}} \propto V$. Surprisingly, this simple fact on its own determines that the ideal gas equation of state is still obeyed! Convince yourself of this by going back over the derivation of the equation of state for such a gas outlined in Section 4.4.

Secondly, you should begin to see how the total heat capacity comes out in the appropriate regime (as shown in Fig. 5.1). Similarly, we can find the formula for an adiabat. For example, we have shown previously for a monatomic gas that the formula for the adiabat was $PV^{5/3} = $ const. Let us now consider a diatomic gas at room temperature, such that $T \gg \theta_{\text{rot}}$ but the temperature is small compared with the vibrational temperature – i.e. we can neglect vibrations. Under these circumstances, to a good approximation:

$$Z_{\text{sp}} = Z_{\text{trans}} \times Z_{\text{rot}} = V \left(\frac{2\pi m k_B T}{h^2}\right)^{3/2} \times \frac{T}{\theta_{\text{rot}}} \propto VT^{5/2} \ .$$

Recall that for a gas of indistinguishable particles

$$S = k_B \ln Z_N + k_B T \frac{\partial \ln Z_N}{\partial T} \ ,$$

where $Z_N = Z_{\text{sp}}^N/N!$ It is thus trivial to show that during an adiabatic expansion ($\Delta S = 0$), $VT^{5/2}$ must be constant. When combined with $PV = RT$ this implies that along an adiabat for a diatomic ideal gas at temperatures such that rotational energy (but not vibrational energy) is important, $PV^{7/5}$ is constant.

5.5 Summary

- Diatomic molecules have rotational and vibrational energy in addition to their translational energy. Taking this into account, the total single-particle partition function can be written:

$$Z_{\text{sp}} = Z_{\text{trans}} \times Z_{\text{rot}} \times Z_{\text{vib}} = V \left(\frac{2\pi m k_B T}{h^2}\right)^{3/2} \times Z_{\text{rot}} \times Z_{\text{vib}} \ .$$

- The rotational partition function is

$$Z_{\text{rot}} = \sum_{J=0}^{\infty} (2J+1) \exp\left[-\hbar^2 J(J+1)/2I k_B T\right] \ .$$

At high temperatures the sum can be replaced by an integral leading to:

$$Z_{\text{rot}} = \frac{1}{\alpha} = \frac{2I k_B T}{\hbar^2} \ ,$$

which leads to a heat capacity of R per mole due to rotations. At low temperatures we can just take the first two terms in the series leading to

$$Z_{\text{rot}} = 1 + 3\exp(-2\alpha) \ .$$

68 Diatomic Ideal Gases

- The vibrational part looks exactly like a simple harmonic oscillator:

$$Z_{\text{vib}} = \exp(-h\nu/2k_BT)\sum_{n=0}^{\infty}\exp(nh\nu/k_BT)$$

$$= \frac{\exp(-h\nu/2k_BT)}{1-\exp(-h\nu/k_BT)} \ .$$

This leads to a heat capacity of

$$C = Nk_B\left(\frac{\theta_{\text{vib}}}{T}\right)^2 \frac{\exp(\theta_{\text{vib}}/T)}{(\exp(\theta_{\text{vib}}/T)-1)^2} \ ,$$

where

$$\theta_{\text{vib}} = \frac{h\nu}{k_B} \ .$$

This heat capacity decays exponentially as $T \ll \theta_{\text{vib}}$, and tends to R per mole as $T \gg \theta_{\text{vib}}$.

- For most diatomic molecules, the separation between the rotational energy levels is much less than k_BT_{room}, but the separation between vibrational levels is much greater than k_BT_{room}. This means that the heat capacity at room temperature is typically $5R/2$ per mole: with $3R/2$ coming from the translational heat capacity and the other R from the rotations. At room temperature the vibrational contribution is small, only coming into play at higher temperatures.

5.6 Problems

1. Show that for a diatomic molecule at a temperature, T, such that $k_BT \gg h\nu$, where ν is its frequency of vibration,

$$Z_{\text{sp}} \propto VT^{7/2} \ , \tag{5.4}$$

and thus that along an isentrope $PV^{9/7}$ is a constant.

2. Given that the internuclear separation in the O_2 molecule is 1.2 Å, calculate its characteristic temperature of rotation.

3. The Einstein temperature (the characteristic temperature of vibration, θ_{vib}) of O_2 is 2200 K. Calculate the percentage contribution of the vibrations to its heat capacity at room temperature.

6
Quantum Statistics

If your experiment needs statistics, you ought to have done a better experiment.
E. Rutherford

6.1 Indistinguishable Particles and Quantum Statistics

In Chapter 4 we learnt that in a gas of like particles we cannot distinguish one from another, and it is necessary to use the statistics of indistinguishable particles. We found that this led to the Boltzmann equation once more. However, within this derivation of the Boltzmann equation in Section 4.2 we made an assumption: we assumed that the number of available states was very large compared with the number of particles (i.e. $g_i \gg n_i$). We said

$$\Omega = \prod_i \frac{(n_i + g_i)!}{n_i!(g_i)!} \to \prod_i \frac{g_i^{n_i}}{n_i!} \quad , \quad \text{when} \quad g_i \gg n_i \quad . \tag{6.1}$$

So we assumed that the particles were never competing to get into the same state – the states were extremely sparsely populated. However, we now consider what happens if this is no longer the case. We might imagine that if we cooled our gas, more and more particles would need to be in the lower energy levels, and so the energy levels would start to be more crowded and become less sparsely populated. This is indeed the case. Furthermore, if we consider our gas to be in a fixed volume (and it is the volume that determines the density of states in a given energy interval), then if we increase the particle density that too would increase the occupation number (i.e. the mean number of particles per state). Thus we reason that if we have a cold, dense system, the assumption that the number of available states is very much larger than the number of particles will eventually break down. This is true and is the basis of quantum statistics, though as we shall see, the situation is somewhat more complicated than that simple picture suggests.

In order to solve this problem we must first find out whether or not more than one particle is allowed, in principle, to be in the same state (a question that never arose in the case of sparsely populated states because it was so unlikely). This is a quantum mechanical problem. Consider the simplest case of a gas of just two identical particles. Let the wavefunction

70 Quantum Statistics

of the combined system be $\psi(1,2)$. Now, as the particles are identical, if we interchange them, no physical observable can change, i.e. $|\psi(1,2)|^2 = |\psi(2,1)|^2$. For this to be true, $\psi(2,1)$ can only differ from $\psi(1,2)$ by a phase factor:

$$\psi(1,2) = \exp(i\delta)\psi(2,1) \ .$$

However, it is obvious that if we swapped the two particles twice, we would have to get back to where we started:

$$\psi(1,2) = \exp(i\delta)\psi(2,1) = \exp(2i\delta)\psi(1,2) = \psi(1,2) \ .$$

So $\exp(2i\delta) = 1, \exp(i\delta) = \pm 1$. Thus, if we have a two-particle wavefunction and we exchange the particles, the new wavefunction must be ± 1 times the old one. Now, here comes another of those annoying bits of physics which, at the level of this book, you will simply have to take on trust. There are two types of particles in the universe from the point of view of statistical mechanics. Those that exchange with the $+1$ factor (i.e. exchanging two particles doesn't alter the two-particle wavefunction) are called 'bosons' and always have integer spin (i.e. 0, 1, 2, etc), whereas the other sort of particles, with the -1 factor upon exchange, are called 'fermions' and have half-integer spin ($\frac{1}{2}, \frac{3}{2}, \frac{5}{2} \ldots$). Thus bosons are said to have *symmetric* wavefunctions upon particle exchange, whereas fermions have *antisymmetric* wavefunctions upon exchange. We have shown this for a two-particle wavefunction, but the point is that this can be demonstrated for the exchange of any two particles in an N-particle system. The fact that there are these two types of particles in the universe has profound consequences when we deal with the statistical mechanics of gases under conditions where we can no longer assume $g_i \gg n_i$.

Indeed, because the particles are in the same box, and we assume the interactions between them are very weak, it is possible to construct the wavefunction for the whole set of particles from linear combinations of single-particle wavefunctions. That is to say, if we deal with the simplest case of just two particles, which can have single-particle wavefunctions of say, ϕ_a and ϕ_b, then one solution to Schrödinger's equation for the two particles together could be

$$\psi(1,2) = \phi_a(1)\phi_b(2) \ , \tag{6.2}$$

i.e. the first particle is in state 'a', and the second is in state 'b'. However, whilst this satisfies the Schrödinger equation, it does not satisfy our criterion that the overall wavefunction is either symmetric (S) or antisymmetric (A) under exchange of particles. Nevertheless, all is not lost, because if $\phi_a(1)\phi_b(2)$ is a solution of Schrödinger's equation, then its interchanged partner $\phi_a(2)\phi_b(1)$ certainly must be as well. It also follows that linear combinations of these two products are also solutions. This then allows us to construct the necessary solutions to Schrödinger's equation that also exhibit the required definite symmetry properties under exchange of particles:

$$\psi_S(1,2) = \frac{1}{\sqrt{2}}[\phi_a(1)\phi_b(2) + \phi_a(2)\phi_b(1)] \quad , \tag{6.3}$$

$$\psi_A(1,2) = \frac{1}{\sqrt{2}}[\phi_a(1)\phi_b(2) - \phi_a(2)\phi_b(1)] \quad . \tag{6.4}$$

Try swapping the labels 1 and 2 in the above equations, and satisfy yourself that they are symmetric and antisymmetric under exchange respectively.

You are probably wondering how this alters the statistics. Well, when we dealt with *distinguishable* particles we didn't worry if more than one of our N particles had the same wavefunction – by definition they could not – either they were intrinsically different particles (like a helium atom as opposed to a xenon atom), or their separate wavefunctions were localized over different, particular regions of space. Also, we didn't worry about it for the perfect gas, where we used indistinguishable statistics, because there we assumed $g_i \gg n_i$, and the particles thus had negligible chance of ever being in the same quantum-mechanical state anyway. However, if we no longer make this assumption, and if the occupation number (which is what we call n_i/g_i) increases, as their wavefunctions are spread over the whole box containing the gas, there is no reason to suspect at present that two of them can't be in the same state. But what do eqns (6.3) and (6.4) imply? What happens if ϕ_a and ϕ_b are the same? It is obvious that if $\phi_a = \phi_b$ then

$$\psi_S(1,2) = \frac{2}{\sqrt{2}}[\phi_a(1)\phi_a(2)] \quad ,$$

$$\psi_A(1,2) = 0 \quad .$$

That is to say, two or more bosons *are* allowed to occupy the same state (the probability is non-zero), but fermions *are not* (the probability is always zero). The fact that no two fermions are allowed to occupy the same state is known as the Pauli exclusion principle. It should be clear that allowing particles to be in the same state or not can alter the statistics. This is the basis of quantum statistics. As we shall show, bosons obey 'Bose–Einstein' statistics, whereas fermions obey 'Fermi–Dirac' statistics. Examples of bosons are ^{4}He atoms, photons and phonons in a crystal, whilst examples of fermions are electrons, ^{3}He atoms (which have odd numbers of spin-1/2 nucleons) and neutrons.

6.2 Bose–Einstein Statistics

When we originally derived the number of ways of arranging indistinguishable particles for the ideal gas, we did not in any way explicitly exclude the

72 Quantum Statistics

possibility of there being more than one particle in the same state. Thus, if we consider eqn (6.1), the result is still valid, as long as we no longer assume $g_i \gg n_i$. That is to say, for bosons

$$\Omega = \prod_i \frac{(n_i + g_i)!}{n_i! g_i!} \quad . \tag{6.5}$$

By now you should be familiar with the set of steps by which we calculate the distribution function from Ω. We maximize $\ln \Omega$ subject to the constraint that there is a fixed number of particles (note that for photons and phonons the number of particles is not fixed - we will return to this point in a moment) and also subject to the constraint that the energy is constant. So, employing Stirling's theorem:

$$\ln \Omega = \sum_i [(n_i + g_i) \ln(n_i + g_i) - n_i \ln n_i - g_i \ln g_i] \quad ,$$

and therefore setting $d \ln \Omega = 0$ implies

$$d \ln \Omega = \sum_i \ln \left(\frac{n_i + g_i}{n_i} \right) dn_i = 0 \quad ,$$

which, with the Lagrange multipliers for the fixed number of particles and energies, leads to:

$$\ln \left(\frac{n_i + g_i}{n_i} \right) + \alpha - \beta \varepsilon_i = 0 \quad ,$$

which finally leads to the distribution function for bosons:

$$\frac{n_i}{g_i} = \frac{1}{\exp(-\alpha + \beta \varepsilon_i) - 1} \quad . \tag{6.6}$$

There are two very important points to note about this distribution function. The first is its consistency with what we have learnt so far. That is to say, if we *do* have a system where $g_i \gg n_i$, then eqn (6.6) reduces to

$$\frac{n_i}{g_i} = \exp(\alpha - \beta \varepsilon_i) \quad ,$$

which is the Boltzmann distribution once more – so we recover our old result if we use our old assumptions. The second important point is to note that if we do not require the number of particles to remain constant (which, as we shall find, is what happens with photons, for example), then we can set $\alpha = 0$. This is the case for a 'gas' of photons or phonons, which we discuss in more detail in Chapter 8. Before that, however, let us derive the distribution function for fermions.

6.3 Fermi–Dirac Statistics

Having derived the distribution function for bosons, let us now consider the statistics of fermions. Once again we consider the situation where we have g_i levels in a bundle, with n_i particles. However, because of the exclusion principle, we know that if there are n_i particles, there *must* be $(g_i - n_i)$ unfilled levels (remember, only one particle can be in each state). So in this case we have g_i objects (the total number of levels), but each of the set of full or unfilled levels are indistinguishable amongst themselves. Therefore the number of ways, Ω_i, of arranging the particles in the ith level is:

$$\Omega_i = \frac{g_i!}{n_i!(g_i - n_i)!} \quad ,$$

and for the whole system of levels

$$\Omega = \prod_i \frac{g_i!}{n_i!(g_i - n_i)!} \quad .$$

Back to the sausage machine (take logarithms, simplify using Stirling's theorem, and find the maximum subject to the constraints of constant particle number and energy) we find:

$$\frac{n_i}{g_i} = \frac{1}{\exp(-\alpha + \beta \varepsilon_i) + 1} \quad ,$$

where $\beta = 1/k_\mathrm{B}T$ as usual. Notice that this looks very similar to the Bose–Einstein distribution – the difference is a '+' in the denominator rather than a '-'. Don't be fooled however – this is a non-trivial difference, as we shall discover shortly.

6.4 More on the Quantum Distribution Functions

You will recall that earlier on in Section 4.2 we indicated that the partition function for a collection of indistinguishable particles differed from the distinguishable case by a factor of $N!$ The argument for this was okay, but not particularly general. Also, we have now introduced two other concepts for dealing with quantum statistics, the idea of fermions and bosons.

Referring back to eqn (6.2) let us for the moment consider free particles and write in plane wave form

$$\phi_\mathrm{a}(1) = \exp(ik_1 x_1) \quad \text{and} \quad \phi_\mathrm{b}(2) = \exp(ik_2 x_2) \quad .$$

If the particles are indistinguishable (typical quantum particles!) it should be obvious to you that particle 1 could be placed in principle into state 2 and particle 2 into state 1. This then means that we have also to consider terms in the wavefunction of the form

$$\exp(ik_2x_1) \quad \text{and} \quad \exp(ik_1x_2)$$

(note the interchange of subscripts). Taking all this together we can then suggest that the total wavefunction for two indistinguishable particles should be of the form

$$\psi(1,2) = \exp[i(k_1x_1 + k_2x_2)] \pm \exp[i(k_2x_1 + k_1x_2)] \quad .$$

The + sign refers to bosons (symmetric wavefunction) while the - sign refers to fermions (antisymmetric wavefunction).

Now, the total energy of this system is simple:

$$E = \frac{\hbar^2}{2m}(k_1^2 + k_2^2) \quad ,$$

and this means that we can sketch a k-space diagram for the allowed states within a maximum energy $E_{\max}$. Figure 6.1 just shows a few such states for positive values of k when $E_{\max}$ is small.

You can see that there are just six states (for convenience we have numbered them in the diagram) that can be occupied within a maximum energy of $E_{\max}$. Or are there? Well, there *would* be if all the states were distinguishable, but remember that we are talking about indistinguishable

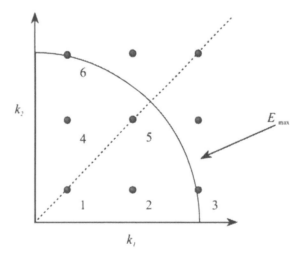

Fig. 6.1 Occupied k-states for a two-particle system. In reality the number of states occupied would normally be huge.

particles and so we have to be careful! Indistinguishability means that we cannot tell the difference between a particle in state labelled 2 and one in state labelled 4 in this diagram, since they have the same magnitude of k and hence the same energy (interchanging k_1 and k_2 must make no difference). You see then that the concept of indistinguishability imposes a mirror symmetry on this diagram, with the dashed line acting as the mirror, with states 1 and 5 lying on the mirror line. When counting states in the indistinguishable case we might consider the states on the diagonal, and those above *or* below the diagonal, but not both those above *and* below it.

Consider now the antisymmetric solution to Schrödinger's equation (corresponding to fermions). Note from our formula for the total wavefunction that in this case the total wavefunction vanishes when $k_1 = k_2$. This is just an example of the Pauli exclusion principle, which does not allow more than one particle in the same state. Therefore, in counting states, the diagonal states carry no weight and so in our example the total number of allowed states is 2 (numbers 2 and 3, or numbers 4 and 6). We can then write that the number of states is $M_{\text{indist,anti}} = 2$ (no weight to the diagonal).

What about bosons? Here the symmetric solution Schrödinger's equation allows the diagonal states to be counted and so we get $M_{\text{indist,sym}} = 4$ (2 diagonal states and 2 non-diagonal states). You can immediately see from this that $M_{\text{indist,anti}} < M_{\text{indist,sym}}$. That's nice because it is consistent with the plus or minus one in the Fermi–Dirac and Bose–Einstein expressions.

Notice the difference from the distinguishable solution where all the states count, $M_{\text{dist}} = 6$. At high temperatures many more states can be populated, so many that we would rather not have to draw them all for you. E_{max} is much bigger and then you can see that the effect of counting or not counting the diagonal states becomes negligible and so we can ignore this fine distinction between diagonal and non-diagonal states. In this limit the two cases, bosons and fermions, become effectively equivalent and our formulae approach the classical Boltzmann result. In this high-temperature regime we can then relate the number of allowed states for the indistinguishable case M_{indist} to that for the distinguishable case M_{dist} by

$$M_{\text{indist}} \approx \frac{M_{\text{dist}}}{2} \ .$$

Let's not forget that our argument so far as been entirely on the basis of only two particles with lots of allowed states. How would our argument change if we had, say, 3 particles? It should be obvious that the number of allowed k-vectors would be 3 i.e. k_1, k_2 and k_3, and so permuting all the possibilities in our total wavefunction formula would give rise to 3! ways of arranging the particles in the allowed states. This idea can then be generalized to the case of N particles, thus:

$$M_{\text{indist}} \approx \frac{M_{\text{dist}}}{N!},$$

and this is why the partition function for indistinguishable particles in the high-temperature limit has the factor of $N!$

6.5 Summary

- If g_i is not exceedingly large compared with n_i, we need to consider whether or not identical particles are allowed to be in the same quantum state.
- Bosons are integer-spin particles. They have symmetric wavefunctions under exchange of particles. There is no restriction on the number of particles in a state. If the number of particles is fixed (e.g. atoms) the distribution function is given by

$$\frac{n_i}{g_i} = \frac{1}{\exp(-\alpha + \varepsilon_i/k_{\text{B}}T) - 1}.$$

If the number of particles is not fixed (photons, phonons) then

$$\frac{n_i}{g_i} = \frac{1}{\exp(\varepsilon_i/k_{\text{B}}T) - 1}.$$

- Fermions are half-integer spin particles. They have antisymmetric wavefunctions under exchange of particles. Only one particle can be in each quantum state (the Pauli exclusion principle). The distribution function is given by

$$\frac{n_i}{g_i} = \frac{1}{\exp(-\alpha + \varepsilon_i/k_{\text{B}}T) + 1}.$$

- Both the Bose–Einstein and Fermi–Dirac distributions tend to the Boltzmann distribution when $n_i \ll g_i$.

6.6 Problems

1. In Section 6.3 we stated that the number of ways of arranging fermions was given by:

$$\Omega = \prod_i \frac{g_i!}{n_i!(g_i - n_i)!}.$$

Use this formula, along with the conservation of particle number and total energy, to derive the Fermi–Dirac distribution function:

$$\frac{n_i}{g_i} = \frac{1}{\exp(-\alpha + \beta\varepsilon_i) + 1}.$$

2. Outline how would you show that $\beta = 1/k_{\text{B}}T$ for a gas of fermions.

7
Electrons in Metals

The electron is not as simple as it looks.
W.L. Bragg

7.1 Fermi–Dirac Statistics: Electrons in Metals

In this chapter we consider an example, in fact the best known example, of Fermi–Dirac statistics: an electron gas. Electrons have a spin of 1/2, and are thus fermions. Now surprisingly, a good approximation for the physics of the conduction electrons in a metal is to treat them as gas particles, ignoring the Coulomb force between them and their interactions with the ionic cores: this is known as the free electron model. The reasons for these assumptions are beyond the scope of this book, and we refer the interested reader to standard texts on solid state physics.[1] For the purposes of the present discussion you will have to take this remarkable fact on trust.

Recall the Fermi–Dirac distribution function assuming closely spaced energy levels:

$$n(\varepsilon)d\varepsilon = \frac{g(\varepsilon)d\varepsilon}{\exp(-\alpha + \varepsilon/k_BT) + 1}$$

Now, the number of electrons in our electron gas *is* fixed, and thus α must be some sort of function of temperature. This can be seen by considering the equation:

$$N = \int_0^\infty \frac{g(\varepsilon)d\varepsilon}{\exp(-\alpha)\exp(\varepsilon/k_BT) + 1} \quad . \tag{7.1}$$

We clearly see that in order for the integral to always have the same value (which it must, as the left-hand side is a constant), then α must be some function of temperature, in order to make the integral 'adjust'. By convention, and for reasons that will become apparent as we proceed, we let

$$\exp(-\alpha) = \exp(-\mu/k_BT) \quad ,$$

where μ is called the chemical potential (itself some function of temperature), and so

---

[1] See for example "*Introduction to Solid State Physics*" by C. Kittel.

78   Electrons in Metals

$$f(\varepsilon) = \frac{n(\varepsilon)}{g(\varepsilon)} = \frac{1}{\exp[(\varepsilon - \mu)/k_\mathrm{B}T] + 1} \ . \tag{7.2}$$

$f(\varepsilon)$ is known as the occupation number, and tells us how many particles there are per spatial quantum state.

Now, in passing, a word about the chemical potential, μ, which is related to α. Recall that in our derivation of the Fermi–Dirac distribution function we had two Lagrange multipliers: α and β. Now β can be shown to be related to temperature by the same sort of arguments as we used for the Boltzmann distribution. Our arguments were based on the zeroth law. That is to say, if two thermodynamic systems are put into contact with one another such that energy can flow from one to the other, energy will flow until equilibrium is reached, at which point the two systems are defined to have the same temperature (i.e. βs). Now if you take a look at α, you will see that it is the Lagrange multiplier related to the number of particles. Thus, the analogous law for the chemical potential can be stated: if two thermodynamic systems are put into contact with one another, such that *particles* can flow from one to the other, particles will flow from one to the other until the chemical potentials of the two systems are the same.

Now let's return to studying eqn (7.2) carefully. Let us consider what happens as we tend towards absolute zero $(T \to 0)$. Careful examination of eqn (7.2) shows that under these conditions μ must be positive, otherwise $n(\varepsilon)/g(\varepsilon)$ would be zero at all energies. Taking this into account we see that the $\exp[(\varepsilon - \mu)/k_\mathrm{B}T]$ factor will look like $\exp(-\infty)$ (i.e. 0) for $\varepsilon < \mu$, and $\exp(+\infty)$, (i.e. ∞), for $\varepsilon > \mu$. Therefore,

$$f(\varepsilon) = 1 \qquad (\varepsilon < \mu) \ , \tag{7.3}$$

$$f(\varepsilon) = 0 \qquad (\varepsilon > \mu) \ . \tag{7.4}$$

Thus we would predict that the number of particles in a particular spatial quantum state is unity up to the energy μ, and is zero thereafter. Actually, if we think about it for a moment, this is exactly what we would expect. At the lowest of temperatures the particles are going to have the lowest energy they possibly can. However, because of the Pauli exclusion principle, they are not allowed in the same states, and therefore start to stack up.

Whilst eqn (7.3) implies that each spatial quantum state up to the energy μ can only be filled by one electron, we must recall that so far we have not taken into account the phenomenon of electron spin. For a given spatial wavefunction, an electron can have two values of spin, and thus there are two overall quantum states for every spatial state. So at low temperatures the first spatial state contains two electrons, the next pair of electrons must go into the next spatial state, and so on – and the particles have to stack up in energy in order to obey the condition that no two are allowed with the same overall wavefunction (spin plus space). We call the value of μ at absolute zero the Fermi energy, denoted by E_F. It should be clear from eqns (7.1), (7.3) and (7.4) that:

$$\begin{aligned} N &= 2 \times \int_0^{E_F} g(\varepsilon)\mathrm{d}\varepsilon \\ &= 2 \times \int_0^{E_F} 4\pi m V \frac{(2m)^{1/2}}{h^3}\varepsilon^{1/2}\mathrm{d}\varepsilon \\ &= \left(\frac{2mE_F}{\hbar^2}\right)^{3/2} \frac{V}{3\pi^2} , \end{aligned}$$

where we have used the normal density of states for free particles, and multiplied by an extra factor of two to take into account that each spatial quantum state of the electrons has two spin wavefunctions associated with it. Therefore,

$$E_F = \frac{\hbar^2}{2m}\left(\frac{3\pi^2 N}{V}\right)^{2/3} .$$

Therefore, the Fermi energy only depends on the concentration N/V of electrons. How big is it for a typical metal? Well, for a monovalent metal, by definition each metal ion gives up one electron. So considering typical metallic densities, $N/V \sim 10^{29} \mathrm{m}^{-3}$, we find that $E_F \sim 8 \times 10^{-19}$ J. Normally we use units of eV for electron energies (i.e. take the result in joules, and divide it by $e \sim 1.6 \times 10^{-19}$), and then the Fermi energy for a typical metal is of order 5 eV.

You should realize that this is a startlingly large energy! In our eV units, $k_B T$ at room temperature is about 1/40 eV (this is a good number to remember, and it is obtained by dividing $k_B T_{\mathrm{room}}$ by e). So, for typical metals, the Fermi energy is of the order of 200 times greater than the energy of a molecule of air at room temperature! The reason that the electrons have such a high energy is, of course, due to the Pauli exclusion principle – as no two electrons are allowed in the same overall state (i.e., as we have said before, no more than two per spatial state), they 'stack up' in energy, two by two, until the topmost ones reach the dizzy heights of the Fermi energy. Nevertheless, it is important to realize that the electrons that have an *energy* of 5 eV, do *not* have a *temperature* corresponding to 5 eV. The reason is that it is not allowable in this case to equate energy with $k_B T$ – that's something we can only do in the classical case where particles obey Boltzmann statistics.

7.2 The Heat Capacity of a Fermi Gas

We have worked out what happens at absolute zero; so now what happens as we heat the system up? Well, to find out about this we need to plot the occupation number (i.e. the number of particles per state – $f(\varepsilon)$) for different temperatures. This is shown in Fig 7.1. Note that when we first heat the system up, most of the electrons, the ones with the lower energies, are not affected. Only the high-energy ones start to occupy still higher states, and the occupation number gets 'rounded' at the edges. Eventually, if we heat up the Fermi gas to a high-enough temperature, the distribution resorts once again to a decaying exponential – i.e. back to the Boltzmann distribution. This occurs when we have so heated up the system that the particles are so spread out in energy that the occupation number is everywhere less than 1/2. That is to say, the electrons are no longer competing fiercely to be in the same quantum state – they are no longer 'stacking up' in energy. We will discuss this transition back to the Boltzmann distribution in much more depth in the next section. For now it is sufficient for you to know that, as a rough guide, the transition takes place when $k_B T \sim E_F$. Note that for a typical metal this corresponds to a temperature of around 50 000 K, by which time the metal has long since melted, vaporized, and turned into a plasma – so in fact it never happens in a metal! However, there are electron gases that do obey classical statistics, and certainly in astrophysical plasmas we can find systems where quantum statistics must be used and others where the classical limit is appropriate.

Furthermore, as we noted previously, when we heat the system up, μ must adjust to satisfy the integral in eqn (7.1). At $T = 0$, $\mu = E_F$, and as the temperature increases, μ decreases slowly. If we study the distribution function

$$f(\varepsilon) = \frac{1}{\exp[(\varepsilon - \mu)/k_B T] + 1} ,$$

we see that $f(\varepsilon) = 1/2$ when $\varepsilon = \mu$. When $k_B T \ll E_F$, μ is still very close to E_F, as is also shown in Fig. 7.1. For a metal at room temperature, $k_B T_{room}$ is less than a percent of E_F, and so taking μ to be the same as E_F is a good approximation. As the system gets hotter, μ decreases, and when it finally passes through zero and becomes negative, $f(\varepsilon)$ is less than 1/2 everywhere (this is one way of describing the transition from quantum to classical statistics, as we shall discuss further in Section 7.3).

Let us consider the case when $k_B T \ll E_F$, i.e. quantum statistics hold and $f(\varepsilon) = 1$ for most energies, only deviating from unity at the highest energies. What is the heat capacity of this system? Working it out exactly is a tedious task, but a rough estimate may be made using the following argument. Consider the distribution function shown in Fig. 7.1. At a temperature T (where $k_B T \ll E_F$), only a fraction of the particles change their energy and the rest remain where they are – this is why the occupation number shown in Fig. 7.1. gets 'rounded' about E_F as the temperature increases. What is this fraction that change energy? Very approximately it must be of order $k_B T/E_F$, as only this fraction are within $k_B T$ of the free available states. Furthermore, for these particular particles that do gain some energy upon heating the system, how much energy do each of

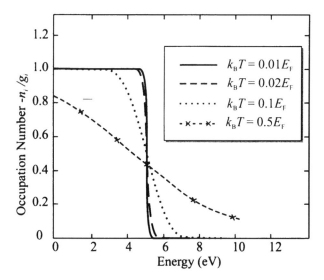

Fig. 7.1 The occupation number, $f(\varepsilon)$, as a function of energy for a Fermi gas, where the density is such that the Fermi energy (i.e. the chemical potential at absolute zero) is 5 eV. The occupation number is shown for four different temperatures. In this plot the temperature dependence of μ has been taken into account and its value can be deduced from the point where the occupation number is 1/2. Note it remains very close to the Fermi energy, decreasing slightly to about 4 eV at $T = 0.5 E_F$.

them gain? Answer: of order $k_B T$. Therefore, the increase in energy of the system, ΔU, is given by:

$$\Delta U \approx N \left(\frac{k_B T}{E_F} \right) k_B T \approx \frac{N k_B^2 T^2}{E_F} \quad .$$

Therefore, the heat capacity of the system is approximated by

$$C \approx \frac{2 N k_B^2 T}{E_F} \quad .$$

We then find that the heat capacity is *proportional* to the temperature. This is borne out by experiment for the electronic contribution to the heat capacity. More detailed arguments than the handwaving one we have employed show that the factor of 2 in the above equation should really be $\pi^2/2$. You should also be aware that the above equation is often written in the form:

$$C = \frac{\pi^2}{2} N k_B \left(\frac{T}{T_F} \right) \quad , \quad \text{where} \quad T_F = \frac{E_F}{k_B} \quad .$$

We see that, to within a factor of order unity, the heat capacity of the Fermi gas is a fraction T/T_F of that of a classical gas (recall that the heat capacity of a monatomic classical gas is $3Nk_B/2$). As we have mentioned already, for a metal T_F is of the order of 50 000 K, and so at room temperature the heat capacity of the electrons in a metal is less than one percent of the heat capacity of a classical gas. Originally, when the free electron model of a metal was put forward, the total heat capacity was a bit of a mystery. It was thought that the heat capacity should be $3Nk_B$ due to the lattice vibrations (as in the high-temperature limit of a 3-D simple harmonic oscillator, for example), and another $3Nk_B/2$ due to the free electrons (i.e. it was assumed that the free electrons would act like a classical gas), and it was not until the advent of quantum statistics and the Pauli exclusion principle that the problem of the missing heat capacity for electrons was resolved.

Now, a word of warning. You will often come across the concept of the 'Fermi temperature', T_F – indeed, we have used it above. It is important to note that this is *not a real temperature* – it is simply E_F/k_B, and T_F is just a convenient notation. Equating $k_B T$ with the energy of the system is something we can only do in classical statistics. This becomes clearer if we reconsider our argument for working out the heat capacity of the Fermi gas. Although we argued that those particles that *did* gain some energy upon heating gained about $k_B T$ each, *most* of the particles did not change their energy at all.

7.3 The Quantum–Classical Transition

We have derived the distribution functions for systems that contain fixed numbers of bosons or fermions:

$$f(\varepsilon) = \frac{n(\varepsilon)}{g(\varepsilon)} = \frac{1}{\exp[(\varepsilon - \mu)/k_B T] \pm 1} .$$

These are the distribution functions for 'quantum statistics' – that is to say, we have taken into account the fact that there may be more than one particle competing to get into the same state. For bosons we are allowed as many particles as we want in the same state, whereas we found that for fermions the Pauli exclusion principle only allowed one particle to have a particular wavefunction (which translates to two particles in each spatial quantum state for electrons, as they can have spin up or down). It should be clear that if the density becomes low, such that $n(\varepsilon) \ll g(\varepsilon)$, i.e. $f(\varepsilon) \ll 1$, the particles are not competing to get into the same state, and we should be able to use the classical formula (Boltzmann statistics) that we derived before. But for $f(\varepsilon) \ll 1$, it must be the case that $\exp(-\alpha + \varepsilon/k_B T) \gg 1$ for all relevant values of ε. Under these conditions we can write the approximate result:

$$f(\varepsilon) = \frac{1}{\exp(-\alpha)} \exp(-\varepsilon/k_B T) = A \exp(-\varepsilon/k_B T)$$

for both Bose–Einstein and Fermi–Dirac gases (you will recognise this already as the Boltzmann distribution). It should hardly be surprising that the classical limit of quantum statistics should get us back to the Boltzmann distribution with which we started. So far, we have simply said that the quantum statistics yields the classical Boltzmann results when $f(\varepsilon)$ is small compared with unity, and the particles are not competing to get into the same energy level. However, there are other ways of looking at this transition between quantum and classical statistics that provide additional insight into the pertinent physics involved.

First, note that the condition for classical statistics to hold is the same as saying $A \exp(-\varepsilon/k_B T) \ll 1$. Furthermore, if this condition is to hold for all energies, it must hold when the exponential factor is largest, i.e. when it is 1 at $\varepsilon = 0$. So finally we see that for classical statistics to be valid, we need $A \ll 1$. Now,

$$N = \int_0^\infty n(\varepsilon) d\varepsilon = \int_0^\infty A g(\varepsilon) \exp(-\varepsilon/k_B T) d\varepsilon = A Z_{\rm sp} \quad ,$$

where $Z_{\rm sp}$ is the single-particle partition function for an ideal gas. We then find that $A \ll 1$ is equivalent to saying:

$$\frac{Z_{\rm sp}}{N} \gg 1 \quad , \quad \text{i.e.} \quad \frac{V}{N}\left(\frac{2\pi m k_B T}{h^2}\right)^{3/2} \gg 1 \quad . \tag{7.5}$$

This enables us to find out in a quantitative way whether classical or quantum statistics are appropriate for fixed numbers of particles, be they bosons or fermions. Notice from the above formula that classical statistics hold when the temperature is high and the density low. This is what we would expect. As the temperature is raised, the particles spread out more over the available quantum states, and there is less chance that they will be competing to get into the same state – so the fact that they are bosons or fermions, having different rules governing whether or not they are allowed in the same quantum state in principle, becomes irrelevant. Clearly, the same argument holds as we reduce the density: we eventually get to a point where the particle density is sufficiently low that they are sufficiently spread out in space that they don't compete to get in the same state.

The above condition, eqn (7.5), gives us a quantitative way of checking whether we can use classical Boltzmann statistics, or whether we need to use the full quantum distribution function. For example, consider a gas of N_2 at STP. Putting the appropriate density and temperature into eqn (7.5) we find $A \sim 2\times 10^{-7}$ (check it!), and we see that this is well into the classical regime. But, on the other hand, electrons in metals have a large value of N/V and typically lead to values of $A \sim 5000$ at room temperature. You can work this out for yourself: e.g. the number density of copper atoms is 8.5×10^{28} m^{-3}. This implies that for the electrons in copper we must use the full Fermi–Dirac distribution function.

Now let us return to study the condition on A a bit more carefully; what does it mean? Well, recall that any particle has an associated quantum mechanical wavelength – the de Broglie wavelength $\lambda_{\rm db}$, related to its momentum, p, by $p = h/\lambda_{\text{db}}$. But we have argued that for small A, the particle obeys classical statistics, and so the energy, $p^2/2m$, must be of order $k_{\text{B}}T$:

$$\frac{p^2}{2m} = \frac{h^2}{2m\lambda_{\text{db}}^2} \sim k_{\text{B}}T \quad ,$$

i.e.

$$\lambda_{\text{db}}^2 \sim \frac{h^2}{2mk_{\text{B}}T} \quad . \tag{7.6}$$

Therefore, by substituting eqn (7.6) into eqn (7.5) we find that, to within a factor of some power of π, our condition to be in the classical regime can be written more simply as

$$\frac{V}{N} \gg (\lambda_{\text{db}}^2)^{3/2} \quad .$$

If the distance between the atoms is d, then since $d = \sqrt[3]{V/N}$

$$d \gg \lambda_{\text{db}} \quad .$$

Let's think about what this elegant and simple expression means: we have classical statistics if the distance between the particles is large compared with their de Broglie wavelength. In many ways this should not be too much of a surprise to us. We have come to expect that quantum phenomena start to rear their heads when we look at distance scales of the order of the de Broglie wavelength. Furthermore, we would expect that the indistinguishability of the particles would start to make a difference when their wavefunctions overlapped significantly, which will happen when $\lambda_{\text{db}} \sim d$.

For electrons there is yet another way we can describe this transition between classical and quantum statistics. Remember that the Fermi energy, equivalent to the chemical potential at absolute zero, is

$$E_{\text{F}} = \frac{\hbar^2}{2m}\left(\frac{3\pi^2 N}{V}\right)^{2/3} \quad . \tag{7.7}$$

If we substitute eqn (7.7) into eqn (7.5) we obtain another way of describing the condition for classical statistics to hold, viz.

$$3\pi^2 \left(\frac{k_{\text{B}}T}{4\pi E_{\text{F}}}\right)^{3/2} \gg 1 \quad ,$$

i.e. we need $k_{\text{B}}T \gg E_{\text{F}}$, which is the same as saying $T \gg T_{\text{F}}$.

In addition, from our earlier discussions of the temperature dependence of μ, we see that the transition from quantum to classical statistics for Fermions can be taken to be the point when μ passes through zero and starts to become negative. When μ is negative, $f(\varepsilon)$ must be less than 1/2 for all energies, and so the particles are not competing heavily to be in the same state.

It is important to remember that once we do get to a condition where classical statistics hold, then for fermions the heat capacity will no longer be proportional to temperature, but will tend to the classical limit of $3Nk_B/2$.

Lastly, a note about nomenclature. In your studies of quantum mechanics you have already come across the word 'degenerate'. An energy level was said to be degenerate if there was more than one wavefunction associated with it. In statistical mechanics, the word degenerate is still used in that way, but it is also applied in a slightly different way to whole systems of particles to mean that they are competing to get into the same state. Thus we say that an electron gas is 'degenerate' if all the particles are competing to get into the same state, and non-degenerate when they are not. By this use of the word, a degenerate system is one where quantum statistics hold, and a non-degenerate system is one which can be well described by classical Boltzmann statistics.

7.4 Summary

- Fermions, which must obey the exclusion principle, obey Fermi–Dirac statistics. For electrons in a metal, this means they 'stack up' in energy. At $T = 0$ the most energetic particles have the Fermi energy:

$$E_F = \frac{\hbar^2}{2m}\left(\frac{3\pi^2 N}{V}\right)^{2/3}.$$

- For electrons in metals, this is a very large energy – of the order of 5 eV. For electrons in metals, room temperature is very much less than the Fermi temperature, and so quantum statistics must be used.
- For fermions that must be described by quantum statistics, the heat capacity of the electrons is proportional to temperature, and a hand-waving argument gives a value of

$$C \approx 2Nk_B \left(\frac{T}{T_F}\right),$$

where $T_F = E_F/k_B$ and is not a real temperature.
- Classical, rather than quantum, statistics can be used to describe the system of a fixed number of particles if:
  (i) the occupation number is low - i.e. $f(\varepsilon) \ll 1$. That is to say, the particles are not competing to get into the same state, and the probability of getting two particles in the same state is negligible.
  (ii) the density is low enough, and the temperature sufficiently high that the following criterion holds:

86 Electrons in Metals

$$\frac{V}{N}\left(\frac{2\pi m k_B T}{h^2}\right)^{3/2} \gg 1 \ .$$

(iii) the distance between the particles is large compared with their de Broglie wavelength, $d \gg \lambda_{db}$.
(iv) in the case of electrons, $k_B T \gg E_F$, i.e. $T \gg T_F$.
(v) for fermions the chemical potential, μ, passes through zero and starts to go negative.

Note that all of the above five statements are totally equivalent. As an aside, also note that for photons $A = \exp(\mu/k_B T) = 1$, and therefore we can never make the above approximation for all energies of radiation. For photons, there will always be a photon energy regime where we have to use the full Bose–Einstein formula (with $A = 1$).

7.5 Problems

1. Show that the mean energy of electrons at absolute zero, $\bar{\varepsilon}$ is $3E_F/5$, where E_F is the Fermi energy. Show also that the ratio of the mean-square-speed to the square of the mean speed is 16/15 (compare this with the ratio found in the case of a 'classical' Maxwell–Boltzmann gas in problem 4 in Section 4.9).

2. Show that the pressure exerted by an electron gas at absolute zero is

$$P = \frac{2NE_F}{5V} \ .$$

3. (i) Show that the equation of state of a perfect gas is still $PV = RT$ even when the gas is heated to such a high temperature that the particles are moving at relativistic speeds. (Hint: what feature of the partition function of the ideal gas determines the above form of the equation of state?)
(ii) Although the equation of state does not alter when the particles in a monatomic ideal gas start to move at relativistic speeds, show that in the formula for an adiabat, $PV^\gamma = $ constant, γ in the relativistic limit is 4/3, rather than 5/3 as in the non-relativistic case.

4. Bill Nellis and his colleagues at Lawrence Livermore National Laboratory in California recently achieved a long-sought after result by producing metallic hydrogen. They achieved this by shock compressing a 0.5 mm thick cryogenic liquid sample of hydrogen, sandwiched between two sapphire plates. The hydrogen was compressed by impacting the sandwiched target with a 'flyer plate' (basically a bullet with a flat surface) travelling at 7 km s^{-1}. The insulator-to-metal transition was seen when the final peak density in the sample was about 0.32×10^6 mole m^{-3}, and the temperature was 3000 K. It is thought that in the metallic phase the liquid comprises H_2^+ ions and electrons. Were the electrons in this liquid metal obeying classical or Fermi–Dirac

statistics? To find out more about this elegant experiment take a look at their results in *Phys. Rev. Lett.* Vol 76, p. 1860 (1996).

5. One possible means of obtaining fusion energy is to implode spherical capsules containing heavy hydrogen by irradiating them with high power lasers. For fusion to occur the implosion core needs to reach electron temperatures of the order of ~1 keV, and the electron density needs to be of the order of 10^{33} m^{-3}. Do we need to use Fermi–Dirac statistics to describe the electrons, or are Maxwell–Boltzmann statistics sufficient?

8
Photons and Phonons

On Monday, when the sun is hot,
I wonder to myself a lot
'Now is it true or is it not,
That what is which and which is what?'
A.A. Milne

8.1 The Photon Gas

Having discussed the statistical mechanics of a gas of fermions, let us now turn our attention to bosons. Recall that bosons are integer-spin particles which have symmetric wavefunctions upon exchange, and there is no restriction on the number occupying the same state.

Before proceeding further it is important to note that bosons themselves come in two varieties. Firstly they can be 'real' particles – like atoms – which obviously come in fixed numbers. Secondly, they can be what we call 'virtual' particles such as photons, which can be created and destroyed, and are variable in number. In this chapter we consider such virtual particles.

Imagine an enclosure which we heat up. The number of photons inside it will increase. We assume that these photons are in thermal equilibrium with the atoms that make up the walls of the enclosure, so that the spectrum of the photons is determined solely by the temperature, rather than the nature of the atoms that make up the walls of the enclosure. Because the number of photons is not a constant, the constraint on the number of particles remaining the same is released, i.e.

$$\sum_i dn_i \neq 0 \quad.$$

If we look back at the procedure whereby we originally obtained the Bose–Einstein distribution in Section 6.2, we see that the Lagrange multiplier α must be zero, and thus $\exp(-\alpha) = 1$. Therefore, for photons, assuming once again that we are dealing with lots of closely spaced levels, we find that the number of particles in the energy interval $d\varepsilon$ is given by:

$$n(\varepsilon)d\varepsilon = \frac{g(\varepsilon)d\varepsilon}{\exp(\varepsilon/k_B T) - 1} \quad, \tag{8.1}$$

where $g(\varepsilon)d\varepsilon$ is once again the density of states. So we must now find the density of states for photons before proceeding further.

8.2 Generalized Derivation of the Density of States

In Section 4.2 we derived the density of states for free particles in a box, and we now wish to do the same for photons and phonons. For the case of free particles, we derived the density of states by looking at the energies, which we found from the solution to Schrödinger's equation. For photons, we can use a similar method based on the solution to Maxwell's equations inside the box. However, it is useful to notice that there is one formula for the density of states which works for both virtual particles, such as photons and phonons, and for real particles. This is outlined below.

If we solve Schrödinger's equation for a free particle in a 3-dimensional infinite square well, we obtain solutions of the form:

$$\psi = A \sin(k_x x) \sin(k_y y) \sin(k_z z) \quad , \tag{8.2}$$

and employing the boundary condition that the wavefunction must be zero at the edges of the box we find

$$k_x = \frac{n_x \pi}{a}, \quad k_y = \frac{n_y \pi}{a}, \quad k_y = \frac{n_z \pi}{a} \quad .$$

Thus,

$$k^2 = \frac{\pi^2}{a^2}(n_x^2 + n_y^2 + n_z^2) \quad . \tag{8.3}$$

Now it is also clear that the solutions to the wave equation for light in a box will also be of a similar form (the solution is slightly different in order to satisfy $\nabla \cdot \boldsymbol{E} = 0$, but this just introduces some cosine rather than sine terms – it doesn't change the thrust of our argument). The form of the equations for light is treated in problem 1 at the end of this chapter. Thus, for free particles and photons the density of states is *the same in 'k-space'*. What is this density of states in k-space? Well, in a similar manner to the method we employed before, we write

$$r^2 = (n_x^2 + n_y^2 + n_z^2) = \frac{k^2 a^2}{\pi^2}$$

and we say that the total number of states with k-vectors of magnitude between 0 and k, $G(k)$, is one-eighth of the volume of the sphere of radius r, i.e.

$$G(k) = \frac{1}{8}\frac{4\pi}{3}\left(\frac{ka}{\pi}\right)^3 = \frac{Vk^3}{6\pi^2} \quad .$$

And therefore the number of states with k-vectors between k and $(k + \mathrm{d}k)$ is

$$g(k)\mathrm{d}k = \frac{\mathrm{d}G(k)}{\mathrm{d}k}\mathrm{d}k = \frac{Vk^2 \mathrm{d}k}{2\pi^2} \quad . \tag{8.4}$$

It is important to understand that eqn (8.4) is applicable to all forms of particles, whether real or virtual.

To get the density of states for photons, we use the different dispersion relation

$$\varepsilon = \hbar c k$$

to yield

$$g(\varepsilon)\mathrm{d}\varepsilon = 2 \times \frac{V\varepsilon^2 \mathrm{d}\varepsilon}{2\pi^2(\hbar c)^3} .$$

Where the factor of two is to take into account the fact that light has two different polarizations – something that we have hitherto ignored in our analysis. In practice it is more normal to deal with the density of states in frequency space:

$$g(\omega)\mathrm{d}\omega = 2 \times \frac{V\omega^2 \mathrm{d}\omega}{2\pi^2 c^3} . \qquad (8.5)$$

To show that eqn (8.4) also works for particles, we rederive the density of states in energy space for particles, i.e. eqn (4.6). To do this we set

$$\varepsilon = \frac{\hbar^2 k^2}{2m}$$

and substitute this dispersion relation into eqn (8.4) to recover

$$g(\varepsilon)\mathrm{d}\varepsilon = 2\pi(2m)^{3/2}\varepsilon^{1/2}\frac{V}{h^3}\mathrm{d}\varepsilon ,$$

which is the same expression as eqn (4.6). So, we have learnt that the density of states in k-space is the same for both real and virtual particles.

8.3 Blackbody Radiation

Having worked out the density of states for photons, let us proceed using our statistical mechanical methods to outline the theory of blackbody radiation – that is to say for a 'gas' of photons in thermal equilibrium. From eqns (8.1) and (8.5) we see that for such a photon gas in thermal equilibrium at a temperature T the number of photons in a given frequency interval is given by

$$n(\omega)\mathrm{d}\omega = 2 \times \frac{V}{2\pi^2 c^3}\omega^2 \times \frac{\mathrm{d}\omega}{\exp(\hbar\omega/k_\mathrm{B}T) - 1} .$$

The number of photons in a given frequency interval is plotted in Fig. 8.1, where the x axis is in units of $x = \hbar\omega/k_\mathrm{B}T$. Note that the peak in the number of photons per unit frequency occurs at $x = 1.59$, i.e. at a frequency of $\omega = 1.59 k_\mathrm{B} T/\hbar$. This can easily be deduced by maximizing $n(\omega)$.

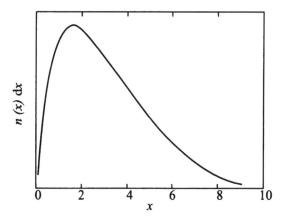

Fig. 8.1 Number of photons per unit frequency interval as a function of frequency for blackbody radiation, where $x = \hbar\omega/k_BT$. The maximum in photon number occurs at $x = 1.59$.

Similarly, the energy per unit frequency interval is

$$U(\omega)d\omega = 2 \times \hbar\omega \times \frac{V}{2\pi^2 c^3}\omega^2 \times \frac{d\omega}{\exp(\hbar\omega/k_BT) - 1} \quad .$$

When plotted as a function of frequency, this has a peak at $\omega = 2.82 k_BT/\hbar$, and the peak of the energy distribution is different from the peak in the photon number because of the ω^3 in the numerator, rather than ω^2. To get the total energy of the radiation in the box we simply integrate the above expression over all frequencies:

$$U = \frac{V\hbar}{\pi^2 c^3} \int_0^\infty \frac{\omega^3 d\omega}{\exp(\hbar\omega/k_BT) - 1} \quad .$$

To evaluate this integral let $x = (\hbar\omega/k_BT)$ and then we write

$$U = \frac{V\hbar}{\pi^2 c^3} \left(\frac{k_BT}{\hbar}\right)^4 \int_0^\infty \frac{x^3 dx}{e^x - 1} \quad .$$

The definite integral has a value of $\pi^4/15$ and so we obtain

$$U = \left(\frac{\pi^2 V k_B^4}{15\hbar^3 c^3}\right) T^4 \quad .$$

Now U/V is the energy density. A standard result in kinetic theory is that the energy flux, η, through a hole of unit area is

$$\eta = \frac{1}{4} c \frac{U}{V} \quad ,$$

where $\bar{c}$ is the mean speed of the particles. And so the total energy flux (i.e. the energy emitted per unit area per unit time by a blackbody) is

$$\eta = \left(\frac{\pi^2 k_B^4}{60\hbar^3 c^2}\right) T^4 = \sigma T^4 \ .$$

This is known as Stefan's law, and σ is Stefan's constant.

8.4 Phonons

These are quantized thermal waves in a solid. They are also bosons and are not fixed in number, and so the same statistics that we have just studied for photons also applies to phonons. Just like photons they obey a wave equation:

$$\nabla^2 \xi = \frac{1}{v^2} \frac{\partial^2 \xi}{\partial t^2} \ ,$$

where ξ is related to the displacement of the atoms, and v is the speed of sound.[1] Also, the boundary conditions effectively give $\xi = 0$ on the surface of a cube, and furthermore $\varepsilon = \hbar\omega$. However, there is one big difference between photons and phonons: in our integral to work out the total energy in the blackbody radiation we integrated over frequencies between 0 and ∞; however, this is not appropriate for the phonons propagating in a solid, as short wavelengths (corresponding to high frequencies), shorter than twice the interatomic spacing, have no meaning. Thus there is a limit on the highest frequency in a solid. This highest frequency is often represented by something called the Debye frequency, ω_D, which we calculate in the following manner. The density of states for phonons is :

$$g(\omega)d\omega = 3 \times \frac{V\omega^2 d\omega}{2\pi^2 v^3} \ ,$$

where the factor of 3 takes into account that there are now 3 polarizations (2 transverse and 1 longitudinal). Now the total number of modes present within the system of N atoms is $3N$ – this is because each ω actually represents an oscillation of a normal mode of the crystal, and for a one-dimensional chain of N atoms there are N normal modes in total, but for a 3-dimensional array there are $3N$. So ω_D can be calculated from:

$$\int_0^{\omega_D} \frac{3V}{2\pi^2 v^3} \omega^2 d\omega = 3N \ ,$$

---

[1] In the Debye theory we assume that the speed of sound is constant for all phonon frequencies. In practice this is not the case, but the Debye theory is generally used at low temperatures, where only low frequency phonons are excited, and where the approximation that they all travel at the same speed is a good one.

Table 8.1 The heat capacity of silver for various temperatures

Temperature (K)	Heat capacity (J mol^{-1} K^{-1})
0.01	6.46×10^{-6}
0.1	6.48×10^{-5}
1.0	8.17×10^{-4}
10.0	0.177
290.0	24.4

$$\frac{V\omega_D^3}{2\pi^2 v^3} = 3N \quad , \tag{8.6}$$

$$U = \int_0^{\omega_D} U(\omega)\mathrm{d}\omega = \frac{3V\hbar}{2\pi^2 v^3} \int_0^{\omega_D} \frac{\omega^3 \mathrm{d}\omega}{\exp(\hbar\omega/k_B T) - 1} \quad . \tag{8.7}$$

As outlined in problem 2 at the end of this chapter, by writing $x = \hbar\omega/k_B T$ you can show that at low temperatures this leads to $U \propto T^4$ and $C \propto T^3$. This is known as the Debye theory of the heat capacity of a solid – i.e. the T^3 dependence at low temperatures. At high temperatures, it is also possible to show that the heat capacity tends to $3Nk_B$, as in the Einstein model discussed in Section 3.3.2. In the Debye model this occurs at temperatures of the order or greater than the so-called Debye temperature, $\theta_D = \hbar\omega_D/k_B$. For most materials the Debye temperature is around room temperature, and at room temperature the heat capacity is therefore close to $3Nk_B$.

Note that in metals, where both phonons and free electrons are present, at low temperatures the total heat capacity will be of the form $\gamma T + \alpha T^3$, where γ and α are constants (recall that the heat capacity of the electrons is linear with temperature at low temperatures, as described in Section 7.2). Thus, as the temperature is lowered, the total heat capacity of a metal changes from the T^3 dependence due to the phonons to a linear dependence due to the electrons. This can be seen in Table. 8.1, where we list the heat capacity of silver at various temperatures. At the two lowest temperatures in the table, the heat capacity is almost exactly linear in temperature. At slightly higher temperatures, say between 1.0 and 10.0 K, the phonons start to dominate and the T^3 term becomes important. At the highest temperature listed, 290 K, note that the heat capacity approaches $3R$ as expected ($R = 8.315$ J mol^{-1} K^{-1}).

8.5 Summary

- The density of states in 'k-space' is the same for both free particles and photons (with an extra factor of 2 for photons, or 3 for phonons, to take into account the independent polarizations):

$$g(k)\mathrm{d}k = \frac{Vk^2 \mathrm{d}k}{2\pi^2} \quad .$$

The density of states in energy or frequency space can then be easily calculated using the appropriate dispersion relation.

- For blackbody radiation, the number of photons per unit frequency interval has a peak at $\omega = 1.59 k_B T/\hbar$, whereas the energy per unit frequency interval peaks at $\omega = 2.82 k_B T/\hbar$.
- For blackbody radiation, the total energy flux is given by Stefan's law:

$$\eta = \left(\frac{\pi^2 k_B^4}{60 \hbar^3 c^2}\right) T^4 = \sigma T^4 \ .$$

- The heat capacity at low temperatures for a solid is according to the Debye theory given by

$$C_V \propto T^3 \ .$$

8.6 Problems

1. A hollow cubical box with sides of length a has perfectly conducting walls, such that the electric field tangential to the surfaces of the walls must be zero. Show that the system of standing waves for the electric field inside the box

$$E_x = A_x \cos(k_x x) \sin(k_y y) \sin(k_z z) \exp(i\omega t) \ , \quad (8.8)$$
$$E_y = A_y \sin(k_x x) \cos(k_y y) \sin(k_z z) \exp(i\omega t) \ , \quad (8.9)$$
$$E_z = A_z \sin(k_x x) \sin(k_y y) \cos(k_z z) \exp(i\omega t) \ , \quad (8.10)$$

is both a solution to the wave equation for electromagnetic radation

$$\nabla^2 \boldsymbol{E} = \frac{1}{c^2} \frac{\partial^2 \boldsymbol{E}}{\partial t^2} \quad (8.11)$$

and satisfies the boundary conditions for the electric field, provided that $k_x a = n_x \pi$, $k_y a = n_y \pi$, $k_z a = n_z \pi$, where n_x, n_y, and n_z are integers. Thus deduce that eqn (8.3) holds for light as well as particles. Furthermore, demonstrate that this system of equations satisfies the condition $\nabla \cdot \boldsymbol{E} = 0$, given that there are two independent polarizations for light.

2. Starting from eqn (8.7), demonstrate that at low temperatures, i.e. $(T \ll \hbar \omega_D / k_B)$, Debye's theory predicts that the heat capacity of a solid is proportional to T^3.

3. Given that the internal energy U of blackbody radiation in a cavity of volume V at temperature T is given by $U = \alpha V T^4$, where α is independent of both V and T, show by thermodynamic reasoning that

the free energy, F, and the entropy, S, of the radiation must have the forms:

$$F = -\frac{1}{3}\alpha V T^4 + Tf(V) \quad,$$

$$S = \frac{4}{3}\alpha V T^3 - f(V) \quad,$$

where $f(V)$ is an unknown function of V. Assuming S vanishes at $T = 0$ (because there is no radiation) or assuming the third law of thermodynamics:

$$\left(\frac{\partial S}{\partial V}\right)_T \to 0 \quad \text{as} \quad T \to 0$$

show that the radiation pressure is given by

$$P = \frac{1}{3}\frac{U}{V} \quad.$$

4. Outline the steps leading to the formula for the number of photons with frequencies between ω and $\omega + d\omega$ in blackbody radiation at a temperature T:

$$n(\omega)d\omega = 2 \times \frac{V}{2\pi^2 c^3}\omega^2 \times \frac{d\omega}{\exp(\hbar\omega/k_B T) - 1} \quad.$$

Show that $n(\omega)$ has a peak at a frequency given by $\omega = 1.59 k_B T/\hbar$. Furthermore, demonstrate that the energy, $u(\omega)$ peaks at a frequency $\omega = 2.82 k_B T/\hbar$.

5. In 'indirect-drive' laser fusion powerful laser beams are focused through holes onto the inside surface of a small millimeter-scale cylinder. The resultant blackbody radiation inside the cylinder is used to compress a pellet of heavy hydrogen. The two holes through which the laser is focused are 0.8 mm in diameter, and radiation can escape through them. The total power in the beams is of the order of 10^{13} W (this is about the electrical power output of the whole planet at any instant!). Assuming all the laser energy is converted to blackbody radiation, estimate the maximum possible equivalent temperature of the blackbody radiation inside the cylinder (a so-called 'hohlraum') in units of eV.

6. In Section 8.3 we derived the distribution function for blackbody radiation assuming them to be bosons with variable number. This question uses a different approach.

If electromagnetic radiation is contained within a cavity of temperature T, in thermal equilibrium with the cavity, the mean occupation number of a single mode of energy $\hbar\omega$ is given by

$$\langle n \rangle = \frac{1}{\exp(\hbar\omega/k_B T) - 1}.$$

Derive this expression from the assumption that the energy in any one mode of the radiation field is quantized in units of $\hbar\omega$.

7. The sun radiates 3.9×10^{26} J s^{-1}. Find the equilibrium temperature of an insulated blackbody facing the sun at the distance of Mars (2.27×10^8 km). Find also the equilibrium temperature of a perfectly conducting black sphere at this distance. The mean surface temperature on the equator of Mars is 250 K. Comment on the results.

8. Sketch the density of states as a function of frequency for 1, 2, and 3-dimensional solids in the Debye approximation. Consider a two-dimensional solid. Show according to the Debye theory that the heat capacity varies as T^2 at low temperature. In graphite it is found that $C_V \propto T^{2.4}$ at low temperature. Explain this behaviour.

9
Bose–Einstein Condensation

Has he attained the seventh degree of concentration?
G.B. Shaw

9.1 Introduction

In Section 8.3 we studied blackbody radation – that is a thermal distribution of photons, which we recall are bosons. However, we noted in that case that a collection of photons need not conserve particle number. We now turn to the question of the statistics of bosons when particle number *is* conserved. For example, we could consider a collection of ^{4}He atoms. As the atoms contain an even number of fermions, their overall spin is integer, and thus they act as bosons, but we certainly can't create or destroy them in the way we can photons, and particle number is conserved.

9.2 The Phenomenon of Bose–Einstein Condensation

Bose–Einstein condensation is a phenomenon that occurs at low temperatures in a collection of a fixed number of identical bosons. Unlike fermions, we recall that we can have as many bosons as we like in a particular quantum state, and thus if it were possible to cool a system of bosons down to close to absolute zero, we would be unsurprised to find them all sitting in the ground state. Such a collection of bosons would be of considerable interest, as we would have a host of particles all in the same quantum state – such a system would surely exhibit quite unique physical properties as it would act 'as one'. The pertinent question to ask is how cold do we need to make the system before an appreciable fraction of the particles are in this lowest energy quantum state?

Now, *if* the particles were acting classically (which they most certainly are not), we would apply the Boltzmann distribution,

$$\frac{n_i}{g_i} = A\exp(-\varepsilon_i/k_\mathrm{B}T) \quad, \tag{9.1}$$

and conclude that a goodly fraction of the particles will be in the ground state at a temperature θ, where $k_\mathrm{B}\theta \approx \Delta\varepsilon$, and where $\Delta\varepsilon$ is the energy difference between the first excited state and the ground state. It is illuminating to calculate what sort of temperature this (erroneous) classical approach predicts for a real system.

98 Bose–Einstein Condensation

Consider a millimetre-scale droplet of liquid ^{4}He. We ignore the interactions between the particles and treat it as a gas (we comment upon this assumption later on). As we have done on several occasions previously, we work out the energies of the quantum states by treating the gas particles as being confined in a 3-D infinite square well, for which the solution to Schrödinger's equation yields

$$\varepsilon_{n_x n_y n_z} = \frac{h^2}{8ma^2}(n_x^2 + n_y^2 + n_z^2) \quad ,$$

where a is the length of the side of the box (i.e. 1 mm in this case). The values of (n_x, n_y, n_z) for the ground state are (111), and for one of the degenerate first excited states (211). Thus the gap in energy, $\Delta\varepsilon$, between the ground and first excited state is

$$\Delta\varepsilon = \frac{h^2}{8ma^2}[(2^2 + 1^2 + 1^2) - (1^2 + 1^2 + 1^2)] = \frac{3h^2}{8ma^2} \quad . \tag{9.2}$$

We know the value of a and we know the mass of a helium atom, so we can calculate this energy difference – the result is of the order of 2.5×10^{-35} J, which corresponds to a temperature θ of just 1.8×10^{-12} K. Thus the classical Boltzmann formula would predict that we need an incredibly low temperature to get a large fraction of the atoms into the ground state. However, we know that bosons *don't* obey the Boltzmann distribution at low temperature, and we need to take into account the full Bose–Einstein distribution:

$$\frac{n_i}{g_i} = \frac{1}{\exp[(\varepsilon_i - \mu)/k_B T] - 1} \quad . \tag{9.3}$$

What does this distribution tell us about the ground state populations?

First of all we ask ourselves what we expect μ to be. The first thing we note is that we expect μ to be less than the ground state energy of the system. If it were not, then we can see from eqn (9.3) that we would end up with a negative occupation number for the ground state, which is clearly unphysical. Let us study how we would expect μ to vary as we reduce the temperature. We know that as we reduce the temperature, the particles are going to start to occupy states of lower energy. In particular, we will get more and more particles in the ground state as the system becomes colder. From eqn (9.3) we see that for the occupation number to increase, μ must itself increase as the temperature drops; that is to say, μ is always less than the ground state energy ε_G, but must increase and tend towards ε_G as the temperature falls.

Recall that the occupation number of the ground state, n_G, is given by

$$n_G = \frac{1}{\exp[(\varepsilon_G - \mu)/k_B T] - 1} \quad .$$

At very low temperatures we expect many, many particles to be in the ground state. Thus the argument of the exponential must be close to zero, and we can use an expansion:

$$n_G = \frac{1}{1 + [(\varepsilon_G - \mu)/k_B T] - 1} = \frac{k_B T}{\varepsilon_G - \mu} = -\frac{k_B T}{\mu'},$$

where $\mu' = \mu - \varepsilon_G$. As μ' is very close to zero at very low temperatures, we see that n_G becomes huge, i.e. a large number or particles are in the ground state.

We now ask ourselves what the population of the excited states looks like once we have reached this temperature at which a good fraction of the particles are in the ground state. For example, let us assume that we have a total of N particles and we are at a low temperature such that half of them are in the ground state. Under such conditions $\mu' = -2k_B T/N$. Therefore, the number of particles in one of the first excited states is given by

$$\begin{aligned}n_{211} &= \frac{1}{\exp[(\varepsilon_{211} - \mu)/k_B T] - 1} \\ &= \frac{1}{\exp[(\varepsilon_{211} - \varepsilon_G - \mu')/k_B T] - 1} \\ &= \frac{1}{\exp[(\Delta\varepsilon)/k_B T]\exp(2/N) - 1} \\ &\approx \frac{1}{\exp[(\Delta\varepsilon)/k_B T] - 1},\end{aligned}$$

where we note that $\exp(2/N) \approx 1$ as N is such a large number.

At this stage it is a good idea to stand back and remind ourselves where we have got to in the argument – because it is certainly a little complicated. Firstly, we know that μ' is negative, and increases towards zero as the temperature drops. At some point we might expect there to be a good chance of getting some of the bosons in the ground state. If the temperature gets so low that, say, half of them are in the ground state a good approximation for μ' is $\mu' = \mu - \varepsilon_G \approx -2k_B T/N$, which is very small indeed compared with the separations between the energy levels. Thus the first excited state has a population

$$n_{211} = \frac{1}{\exp[(\Delta\varepsilon)/k_B T] - 1},$$

where $\Delta\varepsilon = \varepsilon_{211} - \varepsilon_{111}$. Note that we have left out μ' in the above expression as it is so small. Importantly, μ' can also be left out of the expressions for the populations for all of the other excited states. All this will happen when $|\mu'|$ is small compared with $\Delta\varepsilon$. For this to happen we don't need half of the atoms to be in the ground state (we just picked that fraction for the sake of argument) – we just need a large number compared with the

100 Bose–Einstein Condensation

number in the first excited state. We evidently need to find out at what sort of temperature this occurs. Well, if μ' does become small, the number of atoms in the excited states (i.e. not in the ground state) will be

$$N_{\text{ex}} = \int_0^\infty \frac{g(\varepsilon')\mathrm{d}\varepsilon'}{\exp[\varepsilon'/k_\text{B}T] - 1} \quad,$$

where $\varepsilon' = \varepsilon - \varepsilon_\text{G}$. We know the form for the density of states and thus:

$$N_{\text{ex}} = \frac{4\pi mV}{h^3} \int_0^\infty \frac{\sqrt{2m\varepsilon'}\mathrm{d}\varepsilon'}{\exp[\varepsilon'/k_\text{B}T] - 1} \quad. \tag{9.4}$$

Note that the $\sqrt{\varepsilon'}$ factor in the density of states ensures that the ground state is not counted in the integral as required.

The integral can be found in tables [1] and thus we find

$$N_{\text{ex}} = \left(\frac{2\pi m k_\text{B} T}{h^2}\right)^{3/2} 2.612 V \quad. \tag{9.6}$$

We know that as soon as this becomes less than the original total number of atoms, N, then $(N - N_{\text{ex}})$ of the atoms start to go into the ground state. This is the phenomenon of Bose–Einstein condensation – the fact that at a certain transition temperature many, many of the particles start to occupy the lowest energy state, even though the upper energy states are still quite sparsely populated, and the temperature is relatively high. The transition temperature can be found by considering the temperature at which $N_{\text{ex}} = N$ in eqn (9.6):

$$T_{\text{BE}} = \frac{h^2}{2\pi m k_\text{B}} \left(\frac{N}{2.612 V}\right)^{2/3} \quad. \tag{9.7}$$

For liquid helium, one mole occupies 27.6×10^{-6} m^3, and the transition temperature is about 3 K, which is extremely large compared with the splitting between the ground and first excited states, and much, much larger than our erroneous prediction based on classical Boltzmann statistics.

Below the transition temperature the number of atoms in the ground state is

$$n_\text{G} = N - N_{\text{ex}} = N[1 - (T/T_{\text{BE}})^{3/2}] \quad.$$

The number of atoms in the ground state and $\exp[(\varepsilon_\text{G} - \mu)/k_\text{B}T]$, where μ is the chemical potential, are plotted as a function of temperature in Fig.

---

[1]

$$\int_0^\infty \frac{\sqrt{x}\mathrm{d}x}{\exp(x) - 1} = 1.306\sqrt{\pi} \quad. \tag{9.5}$$

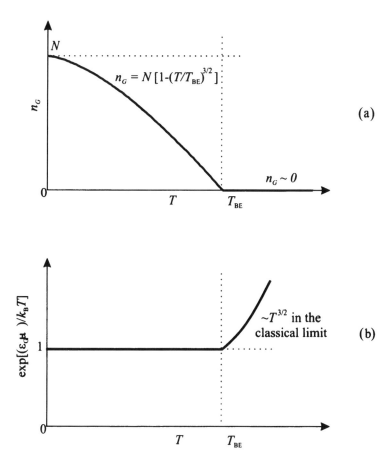

Fig. 9.1 The number of atoms in the ground state and the value of $\exp[(\varepsilon_G - \mu/k_B T]$, where μ is the chemical potential, as a function of temperature for a Bose–Einstein condensate.

9.1. Note that below T_{BE} the number of atoms in the ground state increases dramatically. Below T_{BE} the chemical potential is very close to the ground state, and thus $\exp[(\varepsilon_G - \mu)/k_B T]$ approaches a value of 1. At temperatures well above T_{BE}, we expect classical statistical mechanics to hold and comparing eqn(9.1) with eqn(9.3) we see that under such circumstances $\exp[(\varepsilon_G - \mu)/k_B T] = 1/A$. However, from eqn(4.9) we see that $1/A \simeq Z_{sp}$ which, for an ideal gas, scales as $T^{3/2}$, as shown in Fig. 9.1.

The calculated condensation temperature of 3 K for helium is very close to the actual temperature of 2.17 K at which a transition to a novel state of matter is observed. Below this temperature ^{4}He becomes a 'superfluid' with vanishingly small viscosity. However, despite this, many physicists are reluctant to refer to this as true Bose–Einstein condensation, because in a fluid there are strong forces between the atoms, whereas the theory we have been considering above is for a set of non-interacting particles.

Thus, for many years, indeed since Einstein first predicted the phenomenon in 1925, physicists have tried to get a gas of non-interacting particles to undergo Bose–Einstein condensation. The goal was finally achieved on June 5th 1995. A team of physicists, led by Eric Cornell and Carl Wieman of JILA, Boulder Colorado, USA, finally achieved BEC in a gas of rubidium atoms. The density of the atoms was much less than that of a liquid, and thus (as you can see from eqn (9.7)), the temperature required to achieve BEC was correspondingly lower. The atoms were cooled by sophisticated laser-cooling methods and 'evaporative' cooling in magnetic traps, with a BEC temperature of 1.7×10^{-7} K. Since that time, many groups around the world have made a condensate, and the many atoms in a single quantum state have given rise to a whole host of interesting 'macroscopic' quantum mechanical experiments.

9.3 The Quantum–Classical Transition Revisited

Before we close this chapter, let us take a look once more at eqn (9.7), which tells us the transition temperature for BEC. Recall that in Section 7.3 we considered the temperature and density at which the quantum to classical transition took place in a Fermi gas. Note that the transition temperature occurs at exactly the same place once more, that is to say, when the de-Broglie wavelength of the particle is roughly equal to the inter-particle separation. Equation (9.7) corresponds to $\lambda_{\text{db}} \approx d$, where d is the inter-particle separation.

9.4 Summary

- In a system of a fixed number of bosons, below a certain temperature many, many of the particles condense into the same ground quantum state. This is known as Bose–Einstein condensation. The critical temperature is given by

$$T_{\text{BE}} = \frac{h^2}{2\pi m k_{\text{B}}} \left(\frac{N}{2.612 V}\right)^{2/3} .$$

Below this temperature the ground state atoms are all in the same quantum state, and a whole host of interesting physical effects can be investigated.

- Below the transition temperature the number of atoms n_{G} in the ground state is

$$n_{\text{G}} = N - N_{\text{ex}} = N \left[1 - \left(\frac{T}{T_{\text{BE}}}\right)^{3/2}\right] .$$

9.5 Problems

1. Cornell and Weimann achieved Bose–Einstein condensation with a gas of rubidium atoms, which have an atomic weight of 85.47. The condensation temperature was 1.7×10^{-7} K. What was the number density of the condensate? Given that in this first experiment the total number of atoms in the condensate was only of the order of 2000, to what temperature would the gas need to be cooled to ensure a large number of atoms were in the ground state if classical, rather than quantum, statistics held true?

2. Show that below the condensation temperature, T_{BE}, the heat capacity of a gas obeying Bose–Einstein statistics is given by

$$C_V = 1.93 N k_B \left(\frac{T}{T_{BE}}\right)^{3/2}.$$

(Hint: recall that the chemical potential remains very close to zero below T_{BE}. You will also need to make use of the integral given in B.4.)

10
Ensembles

Will you, won't you, will you, won't you, will you join the dance?
Lewis Carrol (C. Dodgson)

10.1 Introduction

We have to confess at this time to having performed a 'sleight of hand' earlier on when deriving the formulae for the Bose–Einstein and Fermi–Dirac distributions. You will recall that the final formulae included a term, μ, the so-called chemical potential. Now, this appeared in a rather arbitrary way, without full justification. The astute reader might also have noticed that we never derived a partition function for Fermi–Dirac or Bose–Einstein systems. Furthermore, for the systems where we did derive a partition function, we assumed that we had a fixed number of particles, N, and a fixed energy, U. A moment's thought should convince you that there might be systems where such an approach will not work. For example, even if we have a fixed number of monatomic ideal gas particles in a box, surely there is some small probability that the particles in the box will have a little more or a little less energy than $3Nk_BT/2$, as there is some chance that the atoms making up the walls of the box might give up a little more or less of the energy that they possess to the gas atoms. Or consider an even more complex case: consider a cubic centimetre of air in the room and draw an imaginary box around it. Even if you know the density and total energy of all of the air in the room, there will be some small fluctuations in the exact number of particles and the exact total energy of the particles within the given cubic centimetre. Evidently we are not going to be able to learn anything about such fluctuations merely by taking an approach that assumes that the total energy and total particle number is always fixed: a more sophisticated approach is necessary.

The more sophisticated approach is the method of ensembles. The method of ensembles is the part of the subject that generally separates the core, simple results of statistical mechanics (on which we have been concentrating) from the more detailed, mathematical (and, of course, formally correct!) areas of study. Our intention in this chapter is simply to introduce you to the concept, to give you a taster that may persuade you to dig deeper than the simple approaches outlined in the rest of the book.

Up until now we have been considering boxes, or sub-systems of N atoms with total energy U. Consider many, many such sub-systems - the whole lot taken together is called an ensemble. There are three types of ensembles that we can imagine:

1. An ensemble of sub-systems, with each sub-system containing a fixed number of atoms and having fixed energy. This is called the *microcanonical ensemble*, and is in fact the ensemble that we have considered up to now in this book, although we have not hitherto referred to it explicitly.
2. Imagine a sub-system in thermal contact with a very large reservoir, so that energy can flow between the sub-system and its reservoir. The total energy of the sub-system+reservoir is kept fixed, but the energy of the *particular* sub-system itself is *not* fixed as it can exchange energy with its reservoir. However, the sub-system still has a fixed number of particles within it. Now consider replicating many, many such sub-systems plus their associated reservoirs – such an ensemble is known as the *canonical ensemble*.
3. Imagine a sub-system again in contact with a very large reservoir, but now such that both energy *and* particles can flow between the sub-system and the reservoir. The total energy and the total number of particles of the sub-system plus reservoir is fixed, but the energy and number of particles in the sub-system itself can vary, as it can exchange energy and particles with the reservoir. Now consider replicating many, many such sub-systems and their associated reservoirs – such an ensemble is known as the *grand canonical ensemble* – which sounds like a great name for a physicist's jazz band.

By considering the grand canonical approach we shall see that we arrive at a much more complete, and more satisfying, derivation of the Bose–Einstein and Fermi–Dirac formulae in which the chemical potential is explicitly introduced at the beginning. Thus, the present chapter can be thought of as a bridge between the simplified form of statistical mechanics used up to now and the more complete theory needed at a higher level.

It is worth pointing out here that these three types of ensembles have a direct correspondence with the notion of systems as defined within the framework of classical macroscopic thermodynamics. These thermodynamic systems come in three flavours:

1. *Isolated system:* one that is not influenced from outside it boundary. You might like to consider whether the universe can be considered to be an example.
2. *Closed system:* one into and out of which no matter is transferred across its boundary, but influenced in other ways by changes outside the boundary. An example would be a gas in a sealed container being heated from outside by thermal conduction through the walls. As the

walls are heated, the gas pressure and temperature rise, but no gas is allowed to enter or escape.

3. *Open system:* one whose boundary can be crossed by matter, such as a crystal growing in solution.

These three systems clearly relate to microcanonical, canonical and grand canonical ensembles, respectively.

10.2 The Chemical Potential

Before we set about redefining the statistical arguments needed to solve this problem, we wish to consider for the moment a particular bit of macroscopic classical thermodynamics. Normally we are accustomed to thinking of the internal energy U as being a function of heat and work, expressed, for instance, by a relationship involving functions of state, S and V, i.e.

$$U = U(S, V) \quad . \tag{10.1}$$

The most common expression of this relationship is

$$dU = TdS - PdV \quad . \tag{10.2}$$

Now this equation is only valid if we define a thermodynamic system in which the amount of material contained is held constant. But suppose now we allow for the possibility that N_i particles of a substance i are added to the system. Clearly, this will affect the internal energy, and so we can rewrite eqn (10.1) by

$$U = U(S, V, N_i) \quad .$$

This functional relationship can then be written as

$$dU = \left(\frac{\partial U}{\partial S}\right)_{V,N_i} dS + \left(\frac{\partial U}{\partial V}\right)_{S,N_i} dV + \sum_i \left(\frac{\partial U}{\partial N_i}\right)_{S,V,N_{j \neq i}} dN_i \quad .$$

By comparison with eqn (10.2), we can then write

$$dU = TdS - PdV + \sum_i \mu_i dN_i \quad , \tag{10.3}$$

where the chemical potential is given by

$$\mu_i = \left(\frac{\partial U}{\partial N_i}\right)_{S,V,N_{j \neq i}} \quad .$$

The same procedure can be used for the other standard thermodynamic potential equations for enthalpy H, Gibbs free energy G, and Helmholtz free energy F, to give the following equivalent definitions of the chemical potential:

$$\mu_i = \left(\frac{\partial U}{\partial N_i}\right)_{S,V,N_{j\neq i}}$$

$$= \left(\frac{\partial H}{\partial N_i}\right)_{S,P,N_{j\neq i}}$$

$$= \left(\frac{\partial G}{\partial N_i}\right)_{T,P,N_{j\neq i}}$$

$$= \left(\frac{\partial F}{\partial N_i}\right)_{T,V,N_{j\neq i}}.$$

This shows that the chemical potential is a quantity that describes how the various thermodynamic potentials change when particles are added to the system under consideration. This is an important result for thermodynamics because it leads further to a clean definition of what is meant by equilibrium. Thus, suppose there is an equilibrium set up between two phases α and β of a material; we expect to find a balance between the effects of particles of the α phase moving to the β phase and vice versa, so that the equilibrium is defined simply by

$$\mu_\alpha = \mu_\beta . \tag{10.4}$$

10.3 Ensembles and Probabilities

We now leave the world of macroscopic thermodynamics and return to the micro-world of statistical assemblies of particles in order to try to link the two by using the grand canonical ensemble – that is, an ensemble of assemblies of sub-systems and reservoirs where both energy and particles can flow between the two. We denote the sub-system in which we are interested by the symbol S, and the larger reservoir by R, as in Fig. 10.1. Taken together they form a very large system O. For example, the sub-system could be a crystal growing out of a solution, whilst the reservoir could be an enormous volume of the solution. It should be re-emphasised that the sub-system S and the reservoir R allow for exchange of thermal energy as well as for numbers of particles.

Our sub-system is a quantum entity. Up to now we have thought of it as having n_1 particles in the single-particle energy level ε_1, n_2 in the single-particle energy level ε_2, and so on. However, as alluded to in Section 2.5, we can also describe the sub-system in terms of its energy levels as a whole. The total energy of the sub-system, E_s, is related to the single particle energies by

$$E_s = n_1\varepsilon_1 + n_2\varepsilon_2 + \ldots = \sum_i n_i\varepsilon_i$$

and the total number of particles in the sub-system is

$$N_s = \sum_i n_i \ .$$

It is important to recognize that this is an equally valid way of representing the quantum state of the sub-system – there is a quantum state of the whole sub-system, s, with energy E_s (if there is another state with the same overall energy, it is clearly degenerate), containing N_s particles.

We now ask ourselves what is the probability of finding our sub-system in a particular quantum state? Well, we know overall that the total large system, comprising our sub-system and the reservoir, *does* have fixed energy U and particle number N. If our sub-system is in a particular quantum state with energy E_s and particle number N_s, then the energy of the reservoir must be $(U - E_s)$, and it must contain $(N - N_s)$ particles.

If we look at the whole ensemble of such sub-systems, what is the probability of finding a sub-system in a particular quantum state? Well, it must be proportional to the number of ways that the reservoir can have energy $(U - E_s)$ and particle number $(N - N_s)$. In order to determine this number of ways, we return to Boltzmann's original postulate that one can express macroscopic entropy in terms of numbers of microstates, i.e.

$$S = k_B \ln \Omega \quad \text{or} \quad \Omega = \exp(S/k_B) \ . \tag{10.5}$$

The number of arrangements, Ω, is in turn proportional to the probability, P, of being in a particular state. Returning to eqn (10.3) and considering

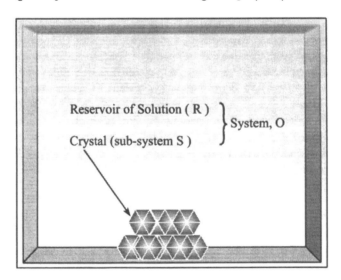

Fig. 10.1 One of the systems, O, in our grand canonical ensemble. Here it comprises a sub-system, S, of a crystal, in contact with a very large reservoir, R, the solution.

now finite changes, we can write that the change in entropy of the reservoir is given by

$$\Delta S = \frac{1}{T}(\Delta U + P\Delta V - \mu \Delta N) \ .$$

We will consider changes at constant volume and so will drop the term in ΔV. Thus, by allowing energy ΔU and particles ΔN to transfer from the reservoir R to the particular sub-system S, the entropy of the reservoir is reduced by the amount

$$\Delta S_R = \frac{1}{T}(\Delta U - \mu \Delta N) \ .$$

This means that the entropy S_R of the reservoir after this process is given by

$$S_R = S_{max} - \frac{1}{T}(\Delta U - \mu \Delta N) \ ,$$

and so, using eqn (10.5), we then get the probability distribution

$$P \propto \exp\left[\frac{S_{max}}{k_B} - \frac{1}{k_B T}(\Delta U - \mu \Delta N)\right]$$

or

$$P \propto A \exp\left[-\frac{1}{k_B T}(\Delta U - \mu \Delta N)\right] \ ,$$

where A is a constant.

Recalling that our sub-system has energy E_s and particle number N_s, we can say that the relative probability of finding such a sub-system in this particular quantum state within the whole ensemble is thus

$$P \propto A \exp\left[-\frac{1}{k_B T}(E_s - N_s \mu)\right] \ . \tag{10.6}$$

To find the *absolute* probability we need a normalizing factor – we normalize by summing over all possible states, s, and all possible particle numbers, and thus obtain

$$P = \frac{\exp\left[-\frac{1}{k_B T}(E_s - N_s \mu)\right]}{\sum_{N=0}^{\infty} \sum_{s(N)} \exp\left[-\frac{1}{k_B T}(E_{s(N)} - N\mu)\right]} \tag{10.7}$$

This normalizing factor is the grand partition function Ξ:

$$\Xi = \sum_{N=0}^{\infty} \sum_{s(N)} \exp\left[-\frac{1}{k_B T}(E_{s(N)} - N\mu)\right] \ .$$

10.4 The Fermi–Dirac and Bose–Einstein Distributions Revisited

We now come to the real magic resulting from adopting the grand canonical formalism – and a both beautiful and subtle piece of physics which will allow us to derive the quantum distribution functions. Up to now in our discussion of the grand canonical ensemble, we have talked in terms of a sub-system in contact with a large reservoir: we gave as an example a crystal in a very large volume of solution. That is to say, we thought in terms of a real piece of 'stuff' (the crystal) in thermal and particle-flow contact with another piece of 'stuff' (the solution). However, let us consider something like a gas of particles (be they bosons or fermions). We know that a box of such particles has particular quantum states and energies (the solutions to Schrödinger's equation for a 3-D infinite square well). Why don't we denote one of these quantum states to be the sub-system, and call all the other quantum states, with their associated particles and thus energy, the reservoir? This is the subtle thought which is the stroke of genius. There is nothing stopping us doing this, as particles and energy can certainly flow into our quantum state (when it becomes occupied) or flow out of it (when a particle transfers to one of the other quantum states), just as particles can flow between the crystal and the solution.

By thinking along these lines we can make use of the probability distribution eqn (10.6) to calculate the mean occupation numbers of quantum states containing either fermions or bosons. If the quantum state we are interested in (the one we call the sub-system) is associated with the energy level ε and it has n particles in it, clearly the energy of this 'sub-system' is $n\varepsilon$. Thus, we can write the mean occupation number of the quantum state as

$$\bar{n} = \frac{\sum nP}{\sum P} = \frac{\sum n \exp[-n(\varepsilon - \mu)/k_BT]}{\sum \exp[-n(\varepsilon - \mu)/k_BT]} \quad , \tag{10.8}$$

where the sum goes over all the possible integer values of n. For convenience, we shall write this as

$$\bar{n} = \frac{\sum nx^n}{\sum x^n} \quad , \tag{10.9}$$

where

$$x = \exp[-(\varepsilon - \mu)/k_BT] \quad .$$

10.5 Fermi–Dirac Distribution Revisited

Consider the application of eqn (10.9) to the case of Fermi–Dirac statistics. As we saw earlier, particles obeying such statistics conform to the Pauli exclusion principle, i.e. it is not possible to have more than one particle in the same quantum state at the same time. This means that the values of n that are possible in eqn (10.9) can only be either 0 or 1. Thus

$$\bar{n} = \frac{0+x}{1+x} = \frac{1}{x^{-1}+1} = \frac{1}{\exp[(\varepsilon-\mu)/k_BT]+1} .$$

10.6 Bose–Einstein Distribution Revisited

In this case, there is no limit to the number of particles that we can place in a particular quantum state with energy ε, and so we now have to sum the complete series in eqn (10.9). Thus we have

$$\bar{n} = \frac{0 + x + 2x^2 + 3x^3 + 4x^4 + \ldots}{1 + x + x^2 + x^3 + x^4 + \ldots} .$$

The denominator is merely a geometric series whose sum, S, is

$$S = (1 + x + x^2 + x^3 + x^4 + \ldots) = (1-x)^{-1} ,$$

and taking the factor x out of the top line we get

$$\bar{n} = x(1 + 2x + 3x^2 + 4x^3 \ldots)/S = (1-x)x(1 + 2x + 3x^2 + 4x^3 \ldots) . \tag{10.10}$$

The remaining series to sum on the right-hand side is a harmonic series. This is just the first derivative of the series S, i.e.

$$1 + 2x + 3x^2 + 4x^3 \ldots = \frac{dS}{dx} = (1-x)^{-2} .$$

Substituting this into eqn (10.10) results in

$$\bar{n} = \frac{(1-x)x}{(1-x)^2} = \frac{1}{x^{-1}-1} = \frac{1}{\exp[(\varepsilon-\mu)/k_BT]-1} .$$

You should now appreciate that both these distribution formulae have appeared with the chemical potential in place naturally and without the need to insert it artificially at the end. This is a very impressive demonstration of the power of using the full treatment afforded by the grand canonical formalism.

10.7 Summary

- In a more sophisticated approach to statistical mechanics, we use the method of ensembles. In particular, the grand canonical ensemble shows us that if both energy and particles can flow, the probability of finding a sub-system with energy E_s and particle number N_s is

$$P \propto A \exp\left[-\frac{1}{k_BT}(E_s - N_s\mu)\right] ,$$

and the the grand partition function is given by Ξ:

$$\Xi = \sum_{N=0}^{\infty} \sum_{s(N)} \exp\left[-\frac{1}{k_BT}(E_{s(N)} - N\mu)\right] .$$

- If, in a collection of bosons or fermions, we call a particular quantum state the 'sub-system' and the rest of the states the 'reservoir', we can quickly derive both the Bose–Einstein and Fermi–Dirac distributions.

10.8 Problems

1. In a paper entitled 'Intermediate Statistics' (Molecular Physics, Vol. 5, p. 525 (1962)) Guénault and MacDonald consider the hypothetical situation where no more than p particles are allowed to be in a given quantum state. Clearly $p = 1$ corresponds to Fermi–Dirac statistics, while $p = \infty$ corresponds to Bose–Einstein statistics. Using the methods outlined in Sections 10.5 and 10.6, show that for this hypothetical case the mean occupation number is given by

$$\bar{n} = \frac{1}{\exp[(\varepsilon - \mu)/k_B T] - 1} - \frac{(p+1)}{\exp[(p+1)(\varepsilon - \mu)/k_B T] - 1}$$

and that this reduces to the Fermi–Dirac and Bose–Einstein distributions in the appropriate limits.

11
The End is in Sight

Entoutes choses il faut considérer le fin.
(In all matters one must consider the end.)
J. de La Fontaine

11.1 Phase Space

In our study of the translational energy of free particles we found that the density of states in energy space was given by

$$g(\varepsilon)d\varepsilon = \frac{4\pi mV}{h^3}(2m\varepsilon)^{1/2}d\varepsilon$$

and

$$Z_{\text{sp}} = \frac{4\pi mV}{h^3}\int_0^\infty (2m\varepsilon)^{1/2}\exp(-\varepsilon/k_\text{B}T)d\varepsilon \quad .$$

Furthermore, when we considered the same problem for photons, we found that the density of states for both particles and photons could be expressed within the same formula – the density of states in 'k-space':

$$g(k)dk = \frac{Vk^2 dk}{2\pi^2} \quad , \tag{11.1}$$

though we need to remember to multiply this by 2 in the case of photons, because they have two distinct polarizations.

For the sake of argument, suppose we change our coordinates to momentum coordinates; for both photons and particles $p = \hbar k$ and thus we find:

$$g(p)dp = \frac{4\pi V p^2 dp}{h^3} \quad . \tag{11.2}$$

It is of interest to express the partition function of the ideal gas in terms of this density of states in momentum space. Clearly,

$$Z_{sp} = \frac{V}{h^3} \int \exp(-p^2/2mk_BT) 4\pi p^2 dp$$

$$= \frac{V}{h^3} \int_{-\infty}^{\infty}\int_{-\infty}^{\infty}\int_{-\infty}^{\infty} \exp(-\varepsilon/k_BT) dp_x dp_y dp_z$$

$$= \frac{1}{h^3} \int_{-\infty}^{\infty}\int_{-\infty}^{\infty}\int_{-\infty}^{\infty}\int_{-\infty}^{\infty}\int_{-\infty}^{\infty}\int_{-\infty}^{\infty} \exp(-\varepsilon/k_BT) dp_x dp_y dp_z dx dy dz \quad.$$

You can see that we are now integrating over a 6-dimensional space whose axes are p_x, p_y, p_z, x, y, z. This space is referred to as 'phase space'. In order to describe the particle we need to specify six coordinates, three for momentum, and three for position in our 6-dimensional phase space.

Generally, therefore,

$$Z_{sp} = \frac{1}{h^3} \int \exp(-\varepsilon/k_BT) d\tau \quad,$$

where the $d\tau$ refers to an integral over the whole of phase space. From the above equation we see that the density of states in phase space is simply

$$g(\tau)d\tau = \frac{1}{h^3} d\tau \quad.$$

It is important to realize that the last result is true for all systems. For example in radiation theory, neglecting the two polarizations, we have

$$g(\omega)d\omega = \frac{V}{2\pi^2 c^3} \omega^3 d\omega$$

and

$$p = \frac{\hbar\omega}{c} \quad.$$

Thus,

$$g(p)dp = \frac{V}{h^3} 4\pi p^2 dp$$

and

$$g(\tau)d\tau = \frac{1}{h^3} d\tau \quad.$$

Note too that phase space is related to Heisenberg's uncertainty principle:

$$\Delta p_x \Delta x \sim h \quad,$$

and so

$$\Delta p_x \Delta x \Delta p_y \Delta y \Delta p_z \Delta z \sim h^3 \quad.$$

In general, we can say that the classical energy surfaces in phase space (surfaces on which particles move with constant energy, known as ergodic surfaces) which correspond to the energies allowed by quantum theory subdivide phase space into units of h^n where n is the number of spatial coordinates. That is, the density of states in phase space is $1/h^n$, where n represents the number of degrees of freedom. Thus if we have free particles with three translational degrees of freedom, for example, we have

$$\begin{aligned}Z_{\text{sp}} &= \frac{1}{h^3}\int_{-\infty}^{\infty}\int_{-\infty}^{\infty}\int_{-\infty}^{\infty}\int_{-\infty}^{\infty}\int_{-\infty}^{\infty}\int_{-\infty}^{\infty}\exp(-\varepsilon/k_\text{B}T)\mathrm{d}p_x\mathrm{d}p_y\mathrm{d}p_z\mathrm{d}x\mathrm{d}y\mathrm{d}z \\ &= \frac{V}{h^3}\int\int\int\exp(-\varepsilon/k_\text{B}T)\mathrm{d}p_x\mathrm{d}p_y\mathrm{d}p_z \\ &= \frac{V}{h^3}\int\exp\left(\frac{-p_x^2}{2mk_\text{B}T}\right)\mathrm{d}p_x\int\exp\left(\frac{-p_y^2}{2mk_\text{B}T}\right)\mathrm{d}p_y\int\exp\left(\frac{-p_z^2}{2mk_\text{B}T}\right)\mathrm{d}p_z \\ &= \frac{V}{h^3}(2\pi mk_\text{B}T)^{3/2} \quad , \end{aligned}$$

which, you must admit, is an elegant way of deriving the translational single-particle partition function for an ideal gas.

11.2 Equipartition of Energy

Let us consider a system for which classical statistics (rather than Bose–Einstein or Fermi–Dirac) is valid. Now, we will define a quantity, called ξ, on which the energy of the system depends in such a way that

$$\varepsilon_{\text{tot}} = \frac{\xi^2}{2M} + \text{everything else} \quad ,$$

where M is some constant. Importantly, notice that we have not explicitly stated what ξ represents; it could be velocity, or position, or something else entirely – we don't know. Neither have we said what M is – we have just said it is a constant. The only thing we really know is that ξ is some coordinate of the system, and the energy is proportional to the square of it. So what? Well, let us look now at the partition function of this system, taking into account what we have learnt about phase space in the previous section:

$$Z_{\text{sp}} = \frac{1}{h^n}\int \exp(-[\xi^2/M + \text{everything else}]/k_\text{B}T)\mathrm{d}x\ldots\mathrm{d}p\ldots\mathrm{d}\xi$$
$$= \frac{1}{h^{n-1}}\int \exp\left(\frac{-[\text{everything else}]}{k_\text{B}T}\right)\mathrm{d}p\ldots\mathrm{d}x\ldots\frac{1}{h}\int \exp\left(\frac{-\xi^2}{2Mk_\text{B}T}\right)\mathrm{d}\xi \quad .$$
(11.3)

Equation (11.3) can be written as

$$Z_{\text{sp}} = Z(\text{everything else}) \times \frac{1}{h}(2\pi Mk_\text{B}T)^{1/2} \quad .$$

Now, the internal energy, U, is given by

$$U = Nk_\text{B}T^2 \frac{\partial \ln Z_\text{sp}}{\partial T} = U(\text{everything else}) + \frac{Nk_\text{B}T}{2} .$$

This is true for any ξ, even though we did not specify what the coordinate actually was! So this result implies something very important:

- **If coordinates enter the expression for the energy as terms proportional to the square of the coordinate, the mean energy per term is the same for all such terms and equals $k_\text{B}T/2$.**

Expressed more simply, for every term expressed in terms of the squares of the coordinates we get a $k_\text{B}T/2$ contribution to the energy. This is a general expression of Maxwell's law of equipartition of energy, and is valid in the high-temperature, low-density limit, where classical statistics are applicable. Thus we see that translational degrees of freedom correspond to kinetic terms with energies such as

$$\frac{1}{2}mv^2 = \frac{1}{2}m\dot{x}^2 ,$$

i.e. squared terms. Vibrational terms have potential energy $m\omega^2 x^2/2$ as well as kinetic energy, and so on. It is because of the squared coordinates in the energy formulae that Maxwell's equipartition law works.

11.3 Route Map through Statistical Mechanics

Now that we have covered the essential details of statistical mechanics, in conclusion it may be helpful to give you a 'map' through it – so you know where to go depending on the problem. Study this carefully, and make sure you understand the reason for each branch in the road.

In terms of the partition function of the *system*, the thermodynamic quantities for when Boltzmann statistics apply (be they distinguishable or indistinguishable particles) are:

$$U = k_\text{B}T^2 \frac{\partial \ln Z_N}{\partial T}$$

$$S = k_\text{B} \ln Z_N + k_\text{B}T \frac{\partial \ln Z_N}{\partial T}$$

$$F = -k_\text{B}T \ln Z_N$$

The above three equations, in combination with the 'route map', constitute the core of this survival guide to statistical mechanics.

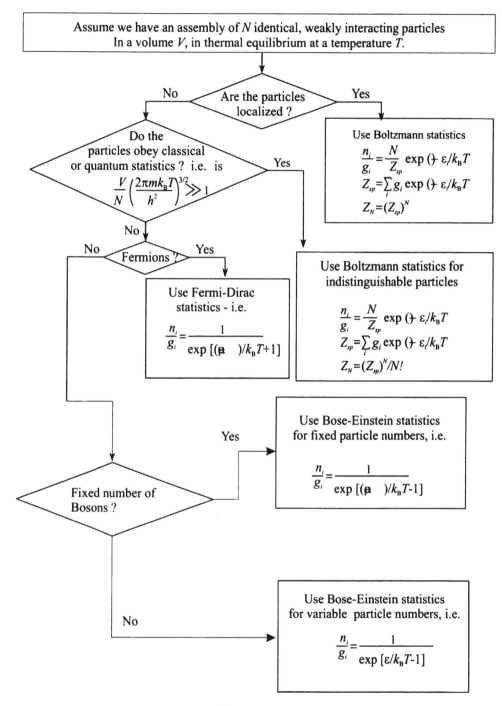

Fig. 11.1
A route map through statistical mechanics.

11.4 Summary

- The density of states in phase space, for a system with n spatial coordinates, is given by

$$g(\tau)d\tau = \frac{1}{h^n}d\tau \;.$$

Thus, in three spatial dimensions each and every h^3 volume of phase space has one quantum state associated with it – a remarkably simple geometric result.

- For a system that obeys classical statistics, the theorem of equipartition of energy states that if coordinates enter the expression for the energy as terms proportional to the square of the coordinate, the mean energy per term is the same for all such terms and equals $k_B T/2$.

11.5 Problems

1. A one-dimensional harmonic oscillator can be described in a phase space which has 2 dimensions with coordinates p and x. Show (a) that the motion in phase space of a classical oscillator with total energy E and frequency ν is an ellipse of area E/ν, (b) that the ellipses which correspond to the energies allowed by quantum mechanics divide this 2-dimensional phase space in equal units of area h.

2. Consider a particle moving freely in a one dimensional 'box' of length a. Show that the area of phase space enclosed between trajectories of energy E_1 and E_2 is given by

$$2a[(2mE_2)^{1/2} - (2mE_1)^{1/2}]$$

and that if E_1 and E_2 are the values corresponding to consecutive quantum states of the particle, this area always equals h.

3. Use the fact that the density of states in phase space is $1/h^n$, where n is the number of spatial coordinates, to derive an expression for the partition function of a 1-D harmonic oscillator in the high-temperature limit. Verify that your expression is the correct high-temperature limit of

$$Z_{\rm sp} = \frac{1}{1 - \exp(-h\nu/k_B T)} \;,$$

where the zero-point energy has been ignored.

4. Given that the number of molecules dn whose generalized coordinates $p_1 \ldots q_1 \ldots$ lie in a small volume $dp_1 \ldots dq_1 \ldots$ is given by

$$\frac{dn}{N} = \frac{\exp(-\varepsilon/k_B T)dp_1 \ldots dq_1 \ldots}{\int \exp(-\varepsilon/k_B T)dp_1 \ldots dq_1 \ldots} \;,$$

derive the principle of equipartition of energy.

A
Worked Answers to Problems

A.1 Chapter 1

1. Our chances are nearly one in 14 million. The chance is

$$\frac{(49 \times 48 \times 47 \times 46 \times 45 \times 44)}{6!} = 13\,983\,816$$

   Why bother? We don't.

2. (i) 1 in 3. There are four 'microstates': - BB, BG, GB, and GG. By telling you that at least one of the children is a boy, Mr. and Mrs. Fatchance have only eliminated the GG microstate. The point is that there are two microstates (BG and GB) in the 'macrostate' of having a boy and a girl. Children are distinguishable particles!
   (ii) Therefore, if they have three children, the total number of available 'microstates' is 8 (2x2x2). They have only eliminated one of them (GGG) by telling you that they have at least one boy – and there is only one microstate associated with the macrostate of having all boys. So the chance of having all boys is now 1 in 7.
   (iii) 1 in $(2^N - 1)$.

3. Yes, the contestant should always change his/her guess to the third, as yet unreferenced, door (i.e. the one that was neither the original guess nor the one opened by the compère). If he/she is stubborn, and keeps the original choice, he/she has a 1 in 3 chance of winning (as at the outset). Swapping every time gives a 2 in 3 chance of winning. The easiest way to see this is to consider what happens if the contestant initially guesses incorrectly: there is a 2 in 3 chance of this occurring, but when it does, clearly the prize lies behind one of the other two doors. The compère is then forced to open the door behind which the prize does not lay, and if the contestant swaps they are now certain to win.

4. Just 23. The easiest way to solve this is to consider the probability for the assembled group of people *not* to share a birthday. This is clearly, for N people,

$$1 \times \frac{364}{365} \times \frac{363}{365} \times \frac{362}{365} \cdots \times \frac{(365-N+1)}{365},$$

and, using a calculator, this is just below 0.5 for $N = 23$.

5. $\ln(10!) = 15.1044126$ whereas $10\ln(10) - 10 = 13.0258509$. There is less than 5% difference for $N = 24$ and less than 1% difference for $N = 91$. Stirling's approximation is quite accurate even for relatively small N.

6. For 6 coins the total number of microstates integrated over all the macrostates is $2^6 = 64$. The number of ways of getting the various macrostates are:
(i) 3H and 3T = $6!/(3! \times 3!) = 20$, (ii) 2H and 4T = $6!/(4! \times 2!) = 15$, and (iii) 1H and 5T = $6!/(5! \times 1!) = 6$.
For 10 coins the total number of microstates is 1024 and the number of microstates in each macrostate obviously goes as 1, 10, 45, 120, 210, 252, 210, 120, 45, 10, 1, for all tails, one head, two heads etc.

A.2 Chapter 2

1. The distribution falls to half of the peak value when:

$$\frac{N!}{(\frac{N}{2}-m)!(\frac{N}{2}+m)!} = \frac{1}{2}\frac{N!}{(\frac{N}{2}!)^2}.$$

Hence, when

$$\left(\frac{N}{2}-m\right)!\left(\frac{N}{2}+m\right)! = 2\left(\frac{N}{2}!\right)^2,$$

$$\left(\frac{N}{2}+m\right)\left(\frac{N}{2}+m-1\right)\cdots\left(\frac{N}{2}+1\right)$$
$$= 2\left(\frac{N}{2}-m+1\right)\left(\frac{N}{2}-m+2\right)\cdots\left(\frac{N}{2}\right).$$

Taking the leading terms, and assuming $N \gg m \gg 1$:

$$\left(\frac{N}{2}\right)^m + [1+2+3\ldots+m]\left(\frac{N}{2}\right)^{m-1}$$
$$\simeq 2\left[\left(\frac{N}{2}\right)^m - [1+2+3\ldots+(m-1)]\left(\frac{N}{2}\right)^{m-1}\right].$$

But

$$1+2+\ldots m \sim m^2/2,$$

leading to

$$m \sim \sqrt{N}.$$

The fractional width therefore goes as $N^{-\frac{1}{2}}$.

2. Let us denote the macrostate with n_1 particles in the first energy level, n_2 in the second etc. by $(n_1, n_2, \ldots)$. The nine possible macrostates, along with their associated number of microstates, are:
   (i) (3, 0, 0, 0, 0, 0, 1) : 4
   (ii) (2, 1, 0, 0, 0, 1, 0) : 12
   (iii)(1, 2, 0, 0, 1, 0, 0) : 12
   (iv)(0, 3, 0, 1, 0, 0, 0) : 4
   (v) (2, 0, 0, 2, 0, 0, 0) : 6
   (vi) (2, 0, 1, 0, 1, 0, 0): 12
   (vii)(1, 1, 1, 1, 0, 0, 0) : 24
   (viii)(1, 0, 3 , 0, 0, 0, 0) : 4
   (ix) (0, 2, 0, 2, 0, 0, 0) : 6
   The total number of microstates is 84. In this (not very good) example the macrostate with the most microstates, (vii), has a flat distribution. However, the *average* distribution is (1.33, 1.00, 0.71, 0.48, 0.29, 0.15, 0.05) – very close to an exponential. As the system gets larger the distribution of the most probable macrostate tends to the average distribution.

3. From normalization we find
$$N = AZ_{\rm sp} \quad,$$
and thus
$$U = \frac{N}{Z_{\rm sp}} \sum_i g_i \varepsilon_i \exp(-\varepsilon_i/k_{\rm B}T) \quad.$$

And if we differentiate the logarithm of the partition function with respect to temperature we find
$$\frac{\partial \ln Z_{\rm sp}}{\partial T} = \frac{1}{Z_{\rm sp}} \sum_i g_i \frac{\varepsilon_i}{k_{\rm B}T^2} \exp(-\varepsilon_i/k_{\rm B}T) \quad,$$
i.e.
$$Nk_{\rm B}T^2 \frac{\partial \ln Z_{\rm sp}}{\partial T} = \frac{N}{Z_{\rm sp}} \sum_i g_i \varepsilon_i \exp(-\varepsilon_i/k_{\rm B}T) = U \quad.$$

4. Let there be N_1 atoms in the ground state with energy ε_1, and N_2 in the excited state with energy ε_2. Therefore
$$\frac{N_2}{N_1} = \frac{A\exp(-\varepsilon_2/k_{\rm B}T)}{A\exp(-\varepsilon_1/k_{\rm B}T)} = \exp[-(\varepsilon_2 - \varepsilon_1)/k_{\rm B}T] \quad.$$

As $(\varepsilon_2 - \varepsilon_1)$ =1 eV, and room temperature is approximately 293 K, we find the ratio is $\exp(-39.62) = 6.2 \times 10^{-18}$! Very few electrons are excited at room temperature. A figure worth remembering is that the energy in electron volts corresponding to $k_{\rm B}T$ at room temperature is approximately 1/40 eV.

A.3 Chapter 3

1. (i) The energies are 0 and, for $m_J = \pm 1$, $-\Delta + g_J m_J \mu_B B$, i.e. 0, $-\Delta + g_J \mu_B B$, and $-\Delta - g_J \mu_B B$.

   (ii) The single-particle partition function is

   $$Z_{sp} = 1 + \exp([-(-\Delta + g_J \mu_B B)/k_B T] + \exp[-(-\Delta - g_J \mu_B B)/k_B T]$$
   $$= 1 + 2\exp(\Delta/k_B T)\cosh\left(\frac{g_J \mu_B B}{k_B T}\right) \ .$$

   Thus the magnetization is

   $$M = Nk_B T \frac{\partial \ln Z_{sp}}{\partial B}$$
   $$= \frac{2N g_J \mu_B \exp(\Delta/k_B T) \sinh\left(\frac{g_J \mu_B B}{k_B T}\right)}{1 + 2\exp(\Delta/k_B T)\cosh\left(\frac{g_J \mu_B B}{k_B T}\right)} \ .$$

   And in the limit of small B

   $$\chi = \frac{\mu_0 N g_J^2 \mu_B^2}{k_B T} \frac{1}{1 + \frac{1}{2}\exp(-\Delta/k_B T)} \ .$$

   (iii) If $k_B T \gg \Delta$, then $\exp(-\Delta/k_B T) \simeq 1$. Thus,

   $$\chi = \frac{2\mu_0 N g_J^2 \mu_B^2}{3 k_B T} = \frac{2C}{3T} \ ,$$

   and when $k_B T \ll \Delta$, then $\exp(-\Delta/k_B T) \simeq 0$. Thus,

   $$\chi = \frac{\mu_0 N g_J^2 \mu_B^2}{k_B T} = \frac{C}{T} \ .$$

   (iv) From above, the ratio of the susceptibilities at a given temperature is 3:2. Only the $m_J = \pm 1$ states contribute to the magnetization. In the low-temperature case, only these lower two states are populated, in the high-temperature case, about one third of the electrons are in the $m = 0$ state.

2. The heat capacity is given by

   $$C = Nk_B \left(\frac{\theta}{T}\right)^2 \frac{\exp(\theta/T)}{(\exp(\theta/T) - 1)^2} \ , \quad \text{where} \quad \theta = \frac{h\nu}{k_B} \ .$$

   In the high-temperature limit, $T \gg \theta$, $C \to Nk_B$, and in the low-temperature limit, $T \ll \theta$,

   $$C \to Nk_B \left(\frac{\theta}{T}\right)^2 \exp(-\theta/T) \ .$$

3. For the simple harmonic oscillator

$$Z_{sp} = \exp(-h\nu/2k_BT) \sum_{n=0}^{\infty} \exp(-nh\nu/k_BT) \quad ,$$

i.e.

$$\ln Z_{sp} = -\frac{h\nu}{2k_BT} + \ln\left(\sum_{n=0}^{\infty} \exp(-nh\nu/k_BT)\right) \quad . \quad (A.1)$$

When substituted into the formula for the internal energy, i.e. eqn (2.25), it is trivial to show that the first term on the right hand side of eqn (A.1) gives rise to a constant extra energy $Nh\nu/2$, and thus does not contribute to the heat capacity. Also,

$$S = Nk_B \ln Z_{sp} + Nk_BT \frac{\partial \ln Z_{sp}}{\partial T} \quad . \quad (A.2)$$

Now, from eqn (A.1) the zero point energy contribution to $\ln Z_{sp}$ is clearly the first term on the right hand side: $-h\nu/2k_BT$. Substituting this term into eqn (A.2) yields zero contribution to the entropy.

4. If we take the first level to have zero energy, the single-particle partition function for the system with the finite number, n, of equispaced energy levels is

$$Z_{sp} = \sum_{i=0}^{n} \exp(-\theta/T) = \frac{1-\exp(-n\theta/T)}{1-\exp(-\theta/T)} \quad .$$

But, ignoring zero-point energies (which is valid as we did not define the absolute energy scale for our particles anyway), the partition functions of simple harmonic oscillators with characteristic temperatures θ and $n\theta$ respectively are

$$Z\left(\frac{\theta}{T}\right) = \frac{1}{1-\exp(-\theta/T)} \quad \text{and} \quad Z\left(\frac{n\theta}{T}\right) = \frac{1}{1-\exp(-n\theta/T)} \quad .$$

Thus

$$Z_{sp} = \frac{Z(\theta/T)}{Z(n\theta/T)}$$

and as the internal energy (and hence heat capacity) is a function of the logarithm of the partition function we find

$$C = C(\theta/T) - C(n\theta/T) \quad .$$

If $n = 2$, the difference in the two Einstein heat capacities will look like the Schottky anomaly and, for large n, we will get back to the heat capacity of a simple harmonic oscillator. In between, note that the peak in the heat capacity is indicative of the energy level spacing, and its decay as the temperature rises gives information on the number of levels, n.

A.4 Chapter 4

1. We recall the relationships between the partition functions of a system of particles in distinguishable and indistinguishable states respectively with the single-particle partition function:

$$Z_{\text{dist}} = Z_{\text{sp}}^N \quad \text{whereas} \quad Z_{\text{indist}} = \frac{Z_{\text{sp}}^N}{N!} \ .$$

Thus $\ln Z_{\text{indist}}$ is less than $\ln Z_{\text{dist}}$ by $\ln N!$ However, since the magnetization in the indistinguishable case is given by

$$M = k_B T \frac{\partial \ln Z_{\text{indist}}}{\partial B}$$

we see that upon differentiation with respect to B the $\ln N!$ term makes no contribution, and the magnetization, and hence susceptibility, are the same in the two cases.

2. In two dimensions the energy levels of the system are given by

$$\varepsilon_{n_x n_y} = \frac{h^2}{8ma^2}(n_x^2 + n_y^2) \ .$$

$G(\varepsilon)$, the number of states with energies between 0 and ε, is in this case equal to the number of points (defined by the n_x, n_y) in the positive quadrant of a circle of radius R, such that

$$R^2 = \frac{8ma^2 \varepsilon_{n_x n_y}}{h^2} \ .$$

Thus,

$$G(\varepsilon) = \frac{1}{4}\pi R^2 = \frac{2\pi m A \varepsilon}{h^2} \ ,$$

where $A = a^2$ is the area of the box. Therefore, the density of states is

$$g(\varepsilon)d\varepsilon = \frac{dG(\varepsilon)}{d\varepsilon}d\varepsilon = \frac{2\pi m A}{h^2}d\varepsilon \ .$$

The single-particle partition function can now be found:

$$Z_{\text{sp}} = \int_0^\infty g(\varepsilon) \exp(-\varepsilon/k_B T) d\varepsilon$$

$$= \int_0^\infty \frac{2\pi m A}{h^2} \exp(-\varepsilon/k_B T) d\varepsilon$$

$$= \frac{2\pi m A k_B T}{h^2} \ .$$

So the internal energy is

$$U = N k_B T^2 \frac{\partial \ln Z_{\text{sp}}}{\partial T} = N k_B T \ .$$

Compare this with the answer for the three-dimensional case: $(3k_B T/2)$. Our answer strongly suggests that the translational internal energy per particle is $k_B T/2$ per dimension. This is indeed the case, and is treated in more detail in Section 11.2.

3. Let K be a constant of normalization. Then the mean-square-speed, $\overline{c^2}$, is given by:

$$\overline{c^2} = \frac{K \int_0^\infty c^4 \exp(-\alpha c^2) dc}{K \int_0^\infty c^2 \exp(-\alpha c^2) dc},$$

where $\alpha = m/(2k_B T)$. Using the integrals to be found in Appendix B we find

$$\overline{c^2} = \frac{3}{2\alpha} = \frac{3k_B T}{m},$$

and thus the root-mean-square-speed is

$$c_{\text{rms}} = \sqrt{\frac{3k_B T}{m}}.$$

A nitrogen molecule has a molecular weight of approximately 28, with a corresponding c_{rms} of 510 m s^{-1}.

4. Let K be a constant of normalization. Then:

$$\frac{\overline{c^2}}{(\overline{c})^2} = \frac{K \int_0^\infty c^4 \exp(-\alpha c^2) dc}{K \int_0^\infty c^2 \exp(-\alpha c^2) dc} \times \left(\frac{K \int_0^\infty c^2 \exp(-\alpha c^2) dc}{K \int_0^\infty c^3 \exp(-\alpha c^2) dc} \right)^2,$$

$$\frac{\overline{c^2}}{(\overline{c})^2} = \frac{\int_0^\infty c^4 \exp(-\alpha c^2) dc \int_0^\infty c^2 \exp(-\alpha c^2) dc}{\left(\int_0^\infty c^3 \exp(-\alpha c^2) dc \right)^2}.$$

Using the integrals given in Appendix B, we obtain

$$\frac{\overline{c^2}}{(\overline{c})^2} = \frac{3\pi}{8}.$$

A.5 Chapter 5

1. $Z_{\text{trans}} \propto V T^{3/2}$. If $k_B T \gg h\nu$, then both Z_{rot} and Z_{vib} are proportional to T. Thus $Z_{\text{sp}} \propto V T^{7/2}$. Now, as the partition function is proportional to volume, $PV = RT$ still holds, along with $V T^{7/2}$ being constant along an adiabat as $\Delta S = 0$ implies that Z_{sp} is constant; thus $PV^{9/7}$ is constant along the adiabat, as long as we remain in a temperature regime such that $k_B T \gg h\nu$.

2. The moment of inertia for an O_2 molecule is

$$I = \sum mr^2 = 1.9 \times 10^{-46} \text{ kg m}^2.$$

Therefore,

$$\theta_{\text{rot}} = \frac{\hbar^2}{2I k_B} = 2.07 \text{ K}.$$

126 Worked Answers to Problems

3. The vibrational contribution to the heat capacity is

$$C = Nk_B \left(\frac{\theta}{T}\right)^2 \frac{\exp(\theta/T)}{(\exp(\theta/T)-1)^2} \quad, \quad \text{where} \quad \theta = \frac{h\nu}{k_B} \quad.$$

Setting $T = 293$K and $\theta = 2200$K we find $C = Nk_B \times 3.1 \times 10^{-2}$. Therefore, the total heat capacity is very close to $2.5Nk_B$ due to the translational and rotational motion, and the percentage contribution due to vibration is 1.2 percent.

A.6 Chapter 6

1. The number of ways of arranging the particles is

$$\Omega = \prod_i \frac{g_i!}{n_i!(g_i-n_i)!} \quad.$$

Taking logarithms and using Stirling's theorem:

$$d\ln\Omega = \sum_i \ln\left(\frac{g_i-n_i}{n_i}\right) dn_i \quad.$$

Setting $d\ln\Omega = 0$ alongside constant particle number and total energy we find

$$\ln\left(\frac{g_i-n_i}{n_i}\right) - \beta\varepsilon_i + \alpha = 0 \quad.$$

Thus,

$$\frac{n_i}{g_i} = \frac{1}{\exp(-\alpha+\beta\varepsilon_i)+1} \quad.$$

2. To show $\beta = 1/k_B T$ we could put this system in thermal contact with a system obeying classical statistics, so that the two systems had the same β, invoke the zeroth law to say the two systems have the same temperature, and use the method outlined in Section 2.2 to show $\beta = 1/k_B T$ for the classical system.

A.7 Chapter 7

1. We note that the distribution of the energies of the particles is of the form

$$n(\varepsilon)d\varepsilon \propto Ag(\varepsilon)d\varepsilon \simeq A'(\varepsilon)^{1/2}d\varepsilon \quad.$$

Therefore,

$$\bar{\varepsilon} = \frac{A' \int_0^{E_F} (\varepsilon)^{3/2} d\varepsilon}{A' \int_0^{E_F} (\varepsilon)^{1/2} d\varepsilon} = \frac{3 E_F}{5} \quad .$$

Now we need to compare $\overline{v^2}$ with $(\bar{v})^2$. Well,

$$\overline{v^2} = \frac{2\bar{\varepsilon}}{m} = \frac{6 E_F}{5m} \quad .$$

However,

$$\bar{v} = \overline{\sqrt{\frac{2\varepsilon}{m}}} \quad .$$

And,

$$\overline{\sqrt{\varepsilon}} = \frac{\int_0^{E_F} A'' \varepsilon \, d\varepsilon}{\int_0^{E_F} A'' \sqrt{\varepsilon} \, d\varepsilon} = \frac{3}{4} \sqrt{E_F} \quad .$$

Therefore,

$$\overline{\sqrt{\varepsilon}} \times \overline{\sqrt{\varepsilon}} = \frac{(\bar{v})^2 m}{2} = \frac{9 E_F}{16} \quad ,$$

$$(\bar{v})^2 = \frac{18 E_F}{16 m} \quad .$$

And we obtain the result:

$$\frac{\overline{v^2}}{(\bar{v})^2} = \frac{6}{5} \times \frac{16}{18} = \frac{16}{15} \quad .$$

2. At absolute zero

$$P = -\frac{dU}{dV} \quad , \quad \text{where} \quad U = \frac{3 N E_F}{5} \quad .$$

Therefore,

$$P = \frac{1}{5}(3\pi^2)^{2/3} \frac{\hbar^2}{m} \left(\frac{N}{V}\right)^{5/3} = \frac{2 N E_F}{5V} = \frac{2U}{3V} \quad .$$

3. (i) Even though the gas particles are moving at relativistic speeds, the partition function is still proportional to V, and therefore the ideal gas equation of state still holds.

(ii) We now need to work out the temperature dependence of the partition function for the relativistic particles. Recall that the density of states in k-space is

$$y(k)dk = \frac{Vk^2 dk}{2\pi^2} ,$$

and for relativistic particles $\varepsilon = c\hbar k$. Therefore, the partition function is

$$Z_{\rm sp} = \int_0^\infty \frac{V\varepsilon^2 \exp(-\varepsilon/k_B T)}{2\hbar^3 c^3 \pi^2} d\varepsilon = \frac{V}{2\hbar^3 c^3 \pi^2} 2!(k_B T)^3 \propto VT^3 .$$

Therefore, VT^3 is constant during an adiabatic expansion. Along with $PV = RT$ this gives the equation of an adiabat in the relativistic limit as $PV^{4/3} = \text{const}$.

4. A density of 0.32 moles per cm^3 implies a Fermi energy of nearly 12 eV, compared with a thermal energy at 3000 K of about 0.3 eV. Therefore, we need to use Fermi–Dirac statistics.

5. 1 keV corresponds to about 1.16×10^7 K. Recall that the condition for classical statistics to be valid is

$$\frac{V}{N}\left(\frac{2\pi m k_B T}{h^2}\right)^{3/2} \gg 1 .$$

In this case the term on the left-hand side is 0.095, and so Fermi–Dirac statistics are necessary. An alternative way of looking at this is that the Fermi energy at this density is 3615 eV, which exceeds the temperature, and so quantum statistics are clearly necessary.

A.8 Chapter 8

1. Substituting eqns (8.8) – (8.10) into eqn (8.11) we find

$$k^2 = k_x^2 + k_y^2 + k_z^2 = \frac{\omega^2}{c^2} .$$

At $x = (0, a)$ E_y and E_z must be zero. From eqn (8.8) we see that this necessitates $k_x a = n_x \pi$, where n_x is an integer. Similar arguments hold for the other two dimensions by cyclic permutation of the indices. Thus, we deduce that eqn (8.3) also holds for light. Substituting eqns (8.8) – (8.10) into $\nabla \cdot \boldsymbol{E} = 0$ we find

$$(k_x A_x + k_y A_y + k_z A_z)(\sin(k_x x)\sin(k_y y)\sin(k_z z)) = 0 .$$

Thus,

$$(k_x A_x + k_y A_y + k_z A_z) = 0 .$$

Therefore, $\nabla \cdot \boldsymbol{E} = 0$ is satisfied, provided that there are two independent polarizations (i.e. A_x, A_y free to vary, for example).

2. From eqn (8.7):

$$U = \frac{3V\hbar}{2\pi^2 v^3} \int_0^{\omega_D} \frac{\omega^3 d\omega}{\exp(\hbar\omega/k_B T) - 1}.$$

Let $x = \hbar\omega/k_B T$, thus we obtain

$$U = \frac{3V\hbar}{2\pi^2 v^3} \left(\frac{k_B T}{\hbar}\right)^4 \int_0^{x_D} \frac{x^3 dx}{\exp(x) - 1},$$

where we define $x_D = \hbar\omega_D/k_B T$. At very low temperatures compared with the so-called Debye temperature, $\theta_D = \hbar\omega_D/k_B$, the upper limit in the integral, x_D, tends towards infinity. At this point the integral becomes a definite integral, the value of which is $\pi^4/15$. Thus the energy of the system at low temperatures is

$$U = \frac{3V\hbar}{2\pi^2 v^3} \frac{\pi^4}{15},$$

which, given our definition of θ_D and eqn (8.6) leads to a heat capacity at low temperatures ($T \ll \theta_D$) of

$$C = \frac{12\pi^4}{5} \left(\frac{T}{\theta_D}\right)^3 N k_B$$

Note that for many crystals θ_D is of the order of room temperature and at and above such temperatures the heat capacity, as in the Einstein model, is close to $3Nk_B$ – it is only at very low temperatures that the heat capacity falls off as T^3.

3. We recall that

$$dF = -SdT - PdV,$$

Therefore,

$$S = -\left(\frac{\partial F}{\partial T}\right)_V,$$

$$U = F + TS = F - T\left(\frac{\partial F}{\partial T}\right)_V = -T^2 \left(\frac{\partial}{\partial T}\left(\frac{F}{T}\right)\right).$$

So we can write

$$U = -T^2 \left(\frac{\partial}{\partial T}\left(\frac{F}{T}\right)\right) = \alpha V T^4,$$

and thus

$$\frac{F}{T} = \frac{-\alpha V T^3}{3} + f(V) \ ,$$

$$F = \frac{-\alpha V T^4}{3} + T f(V) \quad (A.3)$$

And for the entropy we write

$$S = \frac{(U - F)}{T} \ . \quad (A.4)$$

Substituting $U = \alpha V T^4$ and eqn (A.3) into eqn (A.4) we obtain

$$S = \frac{4\alpha V T^3}{3} - f(V) \ .$$

To derive the radiation pressure we note that

$$P = -\left(\frac{\partial F}{\partial V}\right)_T = \frac{\alpha T^4}{3} - T\left(\frac{\partial f(V)}{\partial V}\right)_T \ .$$

But

$$S = \frac{4\alpha V T^3}{3} - f(V)$$

and thus

$$\left(\frac{\partial S}{\partial V}\right)_T = \frac{4\alpha T^3}{3} - \left(\frac{\partial f(V)}{\partial V}\right)_T \ . \quad (A.5)$$

We see from the third law of thermodynamics that the second term on the right-hand side of eqn (A.5) must be zero, and therefore the radiation pressure must be given by

$$P = \frac{\alpha T^4}{3} = \frac{U}{3V} \ .$$

4. Let $x = \hbar\omega/k_B T$, then for photons

$$n(x) \propto \frac{x^2}{\exp(x) - 1} \ .$$

Differentiating $n(x)$ with respect to x and setting the result equal to 0 we find

$$x^2 \exp(x) - 2x \exp(x) + 2x = 0 \ ,$$

which has the solution $x = 1.59$.

And for the energy,

$$U(x) \propto \frac{x^3}{\exp(x) - 1} .$$

Differentiating $U(x)$ with respect to x and setting the result equal to 0 we find

$$x^3 \exp(x) - 3x^2 \exp(x) + 3x^2 = 0 ,$$

which has the solution $x = 2.82$.

5. The area of the holes, A, is

$$A = 2 \times \pi \times (4 \times 10^{-4})^2 = 10^{-6} \text{m}^2 .$$

The maximum temperature will be when we balance input laser power to the output blackbody radiation (this gives the maximum possible temperature – in practice it is less due to albedo effects and the transfer of energy to hydrodynamic motion of plasma, etc.):

$$\sigma T^4 = 5.67 \times 10^{-8} T^4 = 10^{13}/10^{-6} ,$$

$$T = 314 \text{eV} .$$

6. The radiation field is quantized in units of $\hbar\omega$, so that each mode acts like a simple harmonic oscillator. Ignoring the zero point energy (as this cannot be emitted or absorbed), we know that the total energy of N simple harmonic oscillators is

$$U = \frac{N\hbar\omega}{\exp(\hbar\omega/k_B T) - 1} .$$

We know that the energy levels are equispaced, and thus the mean occupation number must be

$$<n> = \frac{1}{N}\frac{U}{\hbar\omega} = \frac{1}{\exp(\hbar\omega/k_B T) - 1} .$$

7. Now the intensity of the sun's radiation at the radius of Mars is 602.3 W m^{-2}. Assuming Mars acts as a perfect blackbody, all this radiation is absorbed and, in equilibrium, must be re-emitted as Mars' own blackbody radation. From Stefan's law this corresponds to a blackbody temperature of $321K$. However, if we assume the planet is a perfectly conducting sphere, then the emitting area is four times greater than that available for absorption. The resultant blackbody temperature from Stefan's law is then $227K$. Mars is somewhere between the two extremes.

8. The internal energy is given by

$$U = \int_0^{\omega_D} \frac{\hbar\omega g(\omega)d\omega}{\exp(\hbar\omega/k_B T) - 1} \ .$$

For a two-dimensional system:

$$g(\omega)d\omega \propto \omega d\omega \ ,$$

and so

$$U \propto \int_0^{\omega_D} \frac{\omega^2 d\omega}{\exp(\hbar\omega/k_B T) - 1} \ .$$

Let $x = \hbar\omega/k_B T$. As $T \to 0$ the integral can be written as

$$U \propto T^3 \int_0^\infty \frac{x^2 dx}{\exp(x) - 1} \ .$$

The definite integral is a constant, and thus $U \propto T^3$ and $C_V \propto T^2$. Graphite consists of carbon atoms covalently (strongly) bonded in layers with weak van der Waals bonding between the layers. It can therefore be thought of as being intermediate between 2- and 3-dimensional.

A.9 Chapter 9

1. The density at which condensation occurs can be found from eqn (9.6), and is given by

$$\frac{N}{V} = 2.612 \left(\frac{2\pi m k_B T_{BE}}{h^2}\right)^{3/2} \ ,$$

which, for the mass and temperature given, corresponds to a number density of 2.74×10^{19} m^{-3}. As there were only 2000 atoms present, the equivalent length of box that contained the atoms was of the order of 4.2μm long. From eqn (9.2), we find that the difference in energy between the ground state and the first excited state is thus 6.6×10^{-32} J. If we equate this energy with $k_B T$ (as is appropriate for classical statistics), we find that 4.8×10^{-9}K would be the classical temperature at which we would expect a significant fraction of the atoms to be in the ground state.

2. Equation (9.4) tells us the number of particles in excited states, and from this equation we can see that the energy of the system below T_{BE} must be given by

$$U = \frac{4\pi mV}{h^3} \int_0^\infty \frac{\sqrt{2m}\varepsilon^{3/2} d\varepsilon}{\exp[\varepsilon/k_B T] - 1} \ .$$

We let $x = \varepsilon/k_BT$ such that

$$U = \frac{4\pi mV}{h^3}\sqrt{2m}(k_BT)^{5/2} \int_0^\infty \frac{x^{3/2}dx}{\exp[x/k_BT] - 1}$$
$$= \frac{4\pi mV}{h^3}\sqrt{2m}(k_BT)^{5/2} \times 1.78 \quad , \tag{A.6}$$

where the definite integral is given in eqn B.4. From eqn (9.7) we find that

$$V = \frac{1}{2.61}\left(\frac{h^2}{2\pi mk_BT_{BE}}\right)^{3/2} . \tag{A.7}$$

By substituting eqn (A.7) into eqn (A.6) we find that below the transition temperature

$$U = \frac{2}{\sqrt{\pi}}\frac{1.78}{2.61}Nk_B\frac{T^{5/2}}{T_{BE}^{3/2}} . \tag{A.8}$$

Differentiating eqn (A.8) with respect to temperature yields the desired result for the heat capacity:

$$C_V = 1.93Nk_B\left(\frac{T}{T_{BE}}\right)^{3/2} .$$

Note that at $T = T_{BE}$ the heat capacity is $1.93Nk_B$, compared with the value of $1.5Nk_B$ that we would expect for a gas obeying classical statistics. Indeed, the heat capacity peaks at $T = T_{BE}$, and then at higher temperatures falls to the classical value.

A.10 Chapter 10

1. Following the notation of Section 10.4, if we are allowed up to p particles in the same quantum state, we have

$$\bar{n} = \frac{0 + x + 2x^2 + 3x^3 + \ldots + px^p}{1 + x + x^2 + x^3 + \ldots + x^p} = \frac{x}{S}\left(\frac{dS}{dx}\right) , \tag{A.9}$$

where S is the series

$$S = 1 + x + x^2 + x^3 + \ldots + x^p = \frac{1 - x^{p+1}}{1 - x} , \tag{A.10}$$

so that

$$\frac{dS}{dx} = \frac{(1 - x^{p+1}) - (p+1)x^p(1 - x)}{(1 - x)^2} . \tag{A.11}$$

Substituting eqns (A.10) and (A.11) into eqn (A.9) we find, after some algebra, that

$$\bar{n} = \frac{x}{1-x} - \frac{(p+1)x^{p+1}}{1-x^{p+1}} \quad,$$

which, given the definition of x in Section 10.4 can be written as

$$\bar{n} = \frac{1}{\exp[(\varepsilon - \mu)/k_BT] - 1} - \frac{(p+1)}{\exp[(p+1)(\varepsilon - \mu)/k_BT] - 1} \quad,$$

as required. It is trivial to show that the above formula reduces to the Fermi–Dirac distribution for $p = 1$ and the Bose–Einstein distribution for $p = \infty$.

A.11 Chapter 11

1. The solution to the equation of motion of a simple harmonic oscillator is

$$x = A\sin(\omega t) \quad , \quad \frac{p}{m} = \frac{dx}{dt} = \omega A \cos(\omega t) \quad .$$

Therefore,

$$\left(\frac{p}{\omega A m}\right)^2 + \left(\frac{x}{A}\right)^2 = 1 \quad ,$$

which is an ellipse of area $\pi A^2 \omega m$. Now the energy, E, of the simple harmonic oscillator is $m\omega^2 A^2/2$, so the area of the ellipse in phase space is $2\pi E/\omega = E/\nu$. As the quantum mechanical solution for the simple harmonic oscillator is $E_n = (n+1/2)h\nu$, we deduce that the ellipses which correspond to the energies allowed by quantum mechanics divide this 2-dimensional phase space into equal units of area h.

2. From $p = \sqrt{2mE}$ the classical phase space is shown in Fig. A.1: The area between the two regions is therefore

$$2a[(2mE_2)^{1/2} - (2mE_1)^{1/2}] \quad .$$

From the Schrödinger equation for a particle in a box:

$$E_n = \frac{n^2\pi^2\hbar^2}{2ma^2} \quad .$$

Therefore, the area, A, in phase space between successive energy states is

$$A = 2a\left[\left(\frac{n^2\pi^2\hbar^2}{a^2}\right)^{1/2} - \left(\frac{(n-1)^2\pi^2\hbar^2}{a^2}\right)^{1/2}\right] = h \quad .$$

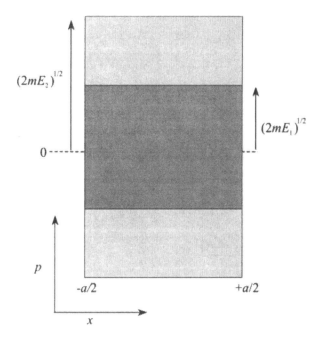

Fig. A.1 Phase space for a 1-D infinite square potential.

3. The total energy of a simple harmonic oscillator is given by

$$\varepsilon = \frac{1}{2}m\omega^2 x^2 + \frac{1}{2}\frac{p^2}{m} \quad .$$

Therefore, we can write the single-particle partition function as

$$Z_{\text{sp}} = \frac{1}{h}\int_{-\infty}^{\infty} \exp(-m\omega^2 x^2/2k_B T)\mathrm{d}x \int_{-\infty}^{\infty} \exp(-p^2/2mk_B T)\mathrm{d}p$$

$$= \frac{1}{h}\left(\frac{2\pi k_B T}{m\omega^2}\right)^{1/2}(2mk_B T\pi)^{1/2} \quad .$$

Thus,

$$Z_{\text{sp}} = \frac{k_B T}{h\nu} \quad ,$$

which is obviously the same as the high-temperature limit of

$$Z_{\text{sp}} = \frac{1}{1 - \exp(-h\nu/k_B T)} \quad .$$

4. We are given the expression for the number of molecules dn whose generalized coordinates $p_1 \ldots q_1 \ldots$ lie in a small volume $\mathrm{d}p_1 \ldots \mathrm{d}q_1 \ldots$:

$$\frac{dn}{N} = \frac{\exp(-\varepsilon/k_B T)dp_1 \ldots dq_1 \ldots}{\int \exp(-\varepsilon/k_B T)dp_1 \ldots dq_1 \ldots}.$$

Assume the energy depends quadratically on the coordinate ζ:

$$\varepsilon = A\zeta^2 + H \ .$$

Thus, the probability of the coordinate ζ lying between ζ and $\zeta + d\zeta$ is

$$P(\zeta)d\zeta = \frac{\exp(-A\zeta^2/k_B T)d\zeta}{\int \exp(-A\zeta^2/k_B T)d\zeta},$$

and the mean value of $A\zeta^2$ is

$$\overline{A\zeta^2} = \frac{\int A\zeta^2 \exp(-A\zeta^2/k_B T)d\zeta}{\int_{\infty}^{\infty} \exp(-A\zeta^2/k_B T)d\zeta},$$

$$\overline{A\zeta^2} = -\frac{\partial}{\partial (1/k_B T)} \ln \int_{\infty}^{\infty} \exp(-A\zeta^2/k_B T)d\zeta,$$

$$\overline{A\zeta^2} = -\frac{\partial}{\partial (1/k_B T)} \left[\ln \sqrt{\frac{\pi k_B T}{A}} \right],$$

$$\overline{A\zeta^2} = \frac{k_B T}{2}.$$

B
Useful Integrals

$$\int_0^\infty x^{2n+1} e^{-a^2 x^2} \, dx = \frac{n!}{2a^{2n+2}} \qquad (B.1)$$

$$\int_{-\infty}^\infty x^{2n} e^{-a^2 x^2} \, dx = \frac{(2n)! \pi^{1/2}}{n!(2a)^{2n} a} \qquad (B.2)$$

$$\int_0^\infty \frac{x^{1/2}}{e^x - 1} \, dx = 2 \cdot 32 \qquad (B.3)$$

$$\int_0^\infty \frac{x^{3/2}}{e^x - 1} \, dx = 1 \cdot 78 \qquad (B.4)$$

$$\int_0^\infty \frac{x^3}{e^x - 1} \, dx = \frac{\pi^4}{15} \qquad (B.5)$$

C
Physical Constants

Constants are given to 4 or 5 significant figures.

Electron rest mass	m_e	9.109×10^{-31} kg
Proton rest mass	M_p	1.6726×10^{-27} kg
Electronic charge	e	1.6022×10^{-19} C
Speed of light in free space	c	2.9979×10^8 m s^{-1}
Planck's constant	h	6.626×10^{-34} J s
$h/2\pi =$	$\hbar$	1.0546×10^{-34} J s
Boltzmann's constant	k_B	1.3807×10^{-23} J K$^-$
Molar gas constant	R	8.315 J mol^{-1} K^{-1}
Avogadro's number	N	6.022×10^{23} mol^{-1}
Standard molar volume		22.414×10^{-3} m^3 m
Bohr magneton	μ_B	9.274×10^{-24} A m^2
Nuclear magneton	μ_N	5.051×10^{-27} A m^2
Stefan's constant	σ	5.671×10^{-8} W m$^-$
Gravitational constant	G	6.673×10^{-11} N m^2
Proton magnetic moment	μ_p	$2.7928\ \mu_N$
Neutron magnetic moment	μ_n	$-1.9130\ \mu_N$
Standard acceleration due to gravity	g	9.807 m s^{-2}
Permeability of free space	μ_0	$4\pi \times 10^{-7}$ H m^{-1}
Permittivity of free space	ε_0	8.854×10^{-12} F m$^-$

Other data and conversion factors

1 angstrom	Å	10^{-10} m
1 pascal	Pa	1 N m^{-2}
1 standard atmosphere		1.0132×10^5 Pa (N m^{-2})
1 electron volt	eV	1.6022×10^{-19} J
	eV/k_B	1.1604×10^4 K
Wavelength of 1 eV photon		1.2399×10^{-6} m

D
Bibliography

As we have explained in the preface and emphasized throughout the text, this book, as its name suggests, is meant to be a survival guide to statistical mechanics, rather than an in-depth, rigorous treatment. Our approach has been motivated by a desire to give students both a grounding and appreciation of the beauty of this subject, without getting bogged down in mathematical detail and subtle nuances. While, in our experience, the majority of working physicists rarely use more than the main results presented within this book, it is clear that the dedicated student will benefit greatly from a study of other texts which present the subject in a more thorough and formal manner. The books listed here are a sample of such treatments that we have found beneficial in the teaching of the subject to second-year physics undergraduates at the University of Oxford.

1. *Statistical Physics*, F. Mandl (Wiley, 1988)
2. *Thermal Physics*, C. Kittel and H. Kroemer (Freeman, 1989)
3. *Statistical Physics*, L.D. Landau and E.M. Lifshitz (Pergamon, 1977)
4. *Probability and Heat*, F. Schlögt (Vieweg, 1989)
5. *Heat and Thermodynamics*, M.W. Zemansky and R.H. Dittman (McGraw-Hill, 1989)
6. *Introduction to Statistical Mechanics and Thermodynamics*, K. Stowe (Wiley, 1984)

Index

adiabatic demagnetization, 55
adiabatic expansion
   of a monatomic ideal gas, 53

blackbody radiation, 90
Boltzmann distribution, 10, 14, 16, 49
Bose–Einstein condensation, 97
   heat capacity of the condenstate, 103, 132
   transition temperature, 100
Bose–Einstein statistics, 71, 111
Brillouin function, 35

canonical ensemble, 105
chemical potential, 77, 106
classical statistics, 25
coins
   macrostates, 3, 5
   microstates, 3, 5
   statistics, 2
Curie's law, 34, 36

de Broglie wavelength, 84, 102
Debye theory of the heat capacity
   of a solid, 93, 94, 129
degeneracy, 22, 85
density of states
   for particles, 49, 90
   for photons, 94, 128
   generalized derivation, 89
   in phase space, 114
diatomic molecules, 62
distinguishable particles, 8
   entropy, 22, 24
   Helmholtz free energy, 22, 24
   internal energy, 22, 24

macrostates, 9, 15
microstates, 10, 15
partition function, 21
   including degeneracy, 23
partition function of a system, 23
Dulong and Petit's law, 40

Einstein model of heat capacity
   of a solid, 38
electrons in metals, 77
   Fermi energy, 79
   heat capacity, 79
ensembles, 104
   canonical, 105
   grand canonical, 105
   microcanonical, 105
entropy
   of a monatomic ideal gas, 53
   of a spin-1/2 paramagnet, 33
   of distinguishable particles, 22, 24
   of indistinguishable particles, 51
   statistical definition, 18
   thermodynamical definition, 19
equipartition of energy, 115
ergodic surfaces, 115
exchange of particles, 71

Fermi energy, 79
Fermi temperature, 82
Fermi–Dirac statistics, 73, 77, 110

Gibbs paradox, 56
grand canonical ensemble, 105

heat capacity

Debye model for a solid, 93, 94, 129
Einstein model for a solid, 38
of a Bose–Einstein condensate, 103, 132
of a gas of fermions, 79
of a simple harmonic oscillator, 38
of a spin-1/2 paramagnet, 32
of an ideal gas, 21
rotational component of a diatomic gas, 63, 64
vibrational component of a diatomic gas, 65
Heisenberg's uncertainty principle, 114
Helmholtz free energy
of a spin-1/2 paramagnet, 34
of distinguishable particles, 22, 24

ideal gas
adiabat, 53
entropy, 53
equation of state, 21, 52
heat capacity, 21
internal energy, 52
rotational heat capacity, 63
rotational partition function, 63
translational partition function, 51
vibrational heat capacity, 64
vibrational partition function, 65
identical particles, 43
indistinguishability of identical particles, 43
indistinguishable particles
entropy, 51
internal energy, 50
statistics, 45
intermediate statistics, 112, 133
internal energy
of a monatomic ideal gas, 52
of a simple harmonic oscillator, 38
of a spin-1/2 paramagnet, 31
of distinguishable particles, 22, 24
of indistinguishable particles, 50
rotational component of a diatomic gas, 64

Lagrange multipliers, 15

macrostates
for coins, 3, 5
for distinguishable particles, 9, 10, 15
Maxwell–Boltzmann distribution of speeds, 56
microcanonical ensemble, 105
microstates
for coins, 3, 5
for distinguishable particles, 10, 15

paramagnet, 29
angular momentum J, 34
magnetization, 35
partition function, 35
spin-1/2, 29
entropy, 33
heat capacity, 32
Helmholtz free energy, 34
internal energy, 31
magnetization, 34
partition function, 30
partition function
for distinguishable particles, 21
of a monatomic ideal gas, 51
of a paramagnet with angular momentum J, 35
of a simple harmonic oscillator, 37
of a spin-1/2 paramagnet, 30
of a system of distinguishable particles, 23
of a system of indistinguishable particles, 51

of distinguishable particles
including degeneracy, 23
rotational component of a diatomic gas, 63
vibrational component of a diatomic gas, 65
Pauli exclusion principle, 71, 79
phase space, 113
phonons, 72, 92
photons, 72, 88
polyatomic molecules, 62

quantum statistics, 69
quantum–classical transition, 82, 102

Sackur–Tetrode equation, 53
Schottky anomaly, 32
Schrödinger's equation
  for a particle in a box, 46
  for two identical particles, 71
simple harmonic oscillator, 10, 36
  heat capacity, 38
  internal energy, 38
  partition function, 37
Stefan's law, 92
Stirling's theorem, 5, 12, 120

temperature
  statistical definition, 20
thermodynamics
  first law, 18
  zeroth law, 18